Exploring Crystalline Glazes

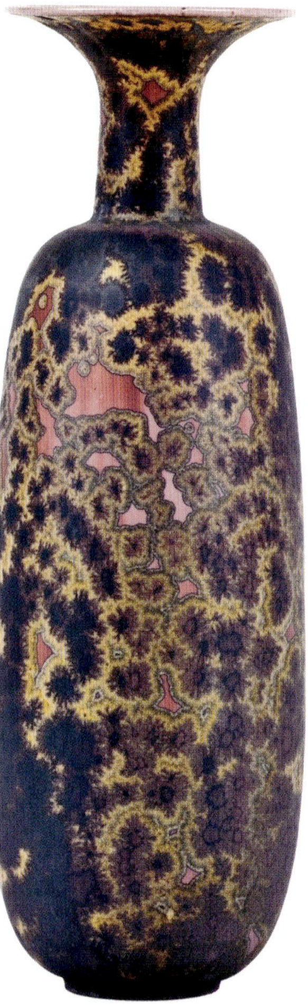

Exploring Crystalline Glazes

Diane Creber

HERBERT PRESS
LONDON • OXFORD • NEW YORK • NEW DELHI • SYDNEY

HERBERT PRESS
Bloomsbury Publishing Plc
50 Bedford Square, London, WC1B 3DP, UK
Bloomsbury Publishing Ireland Limited,
29 Earlsfort Terrace, Dublin 2, D02 AY28, Ireland

First published in Great Britain 2026

A catalogue record for this book is available from the British Library

Library of Congress Cataloguing-in-Publication data has been applied for

ISBN: 978-1-78994-393-1; eBook: 978-1-78994-394-8

2 4 6 8 10 9 7 5 3 1

Designed and typeset by Laura Woussen Design
Printed and bound in China by Toppan Leefung Printing

MIX
Paper | Supporting responsible forestry
FSC® C104723

To find out more about our authors and books visit www.bloomsbury.com and sign up for our newsletters
For product safety related questions contact productsafety@bloomsbury.com

FRONT COVER Top: Silver nitrate added to base glaze then reduction fired. Bottom: Close-up of glaze © Bill Boyd.

BACK COVER Top left: Gold Stuff glaze (see p.116), fired in oxidation. Bottom right: 3 legged bowl. Oxidation fired.

SPINE Reduction-fired bottle, etched in vinegar.

FRONTISPIECE Tall bottle, fired in oxidation then re-fired in reduction.

Contents

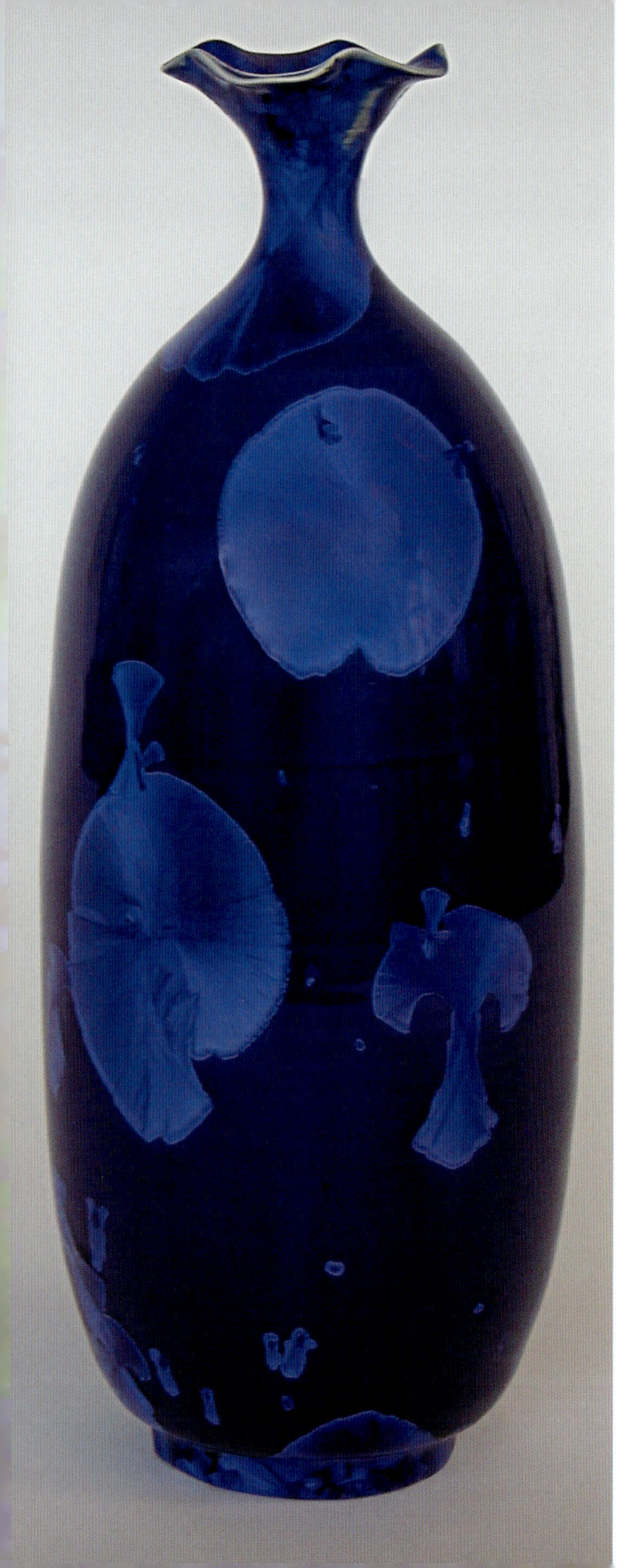

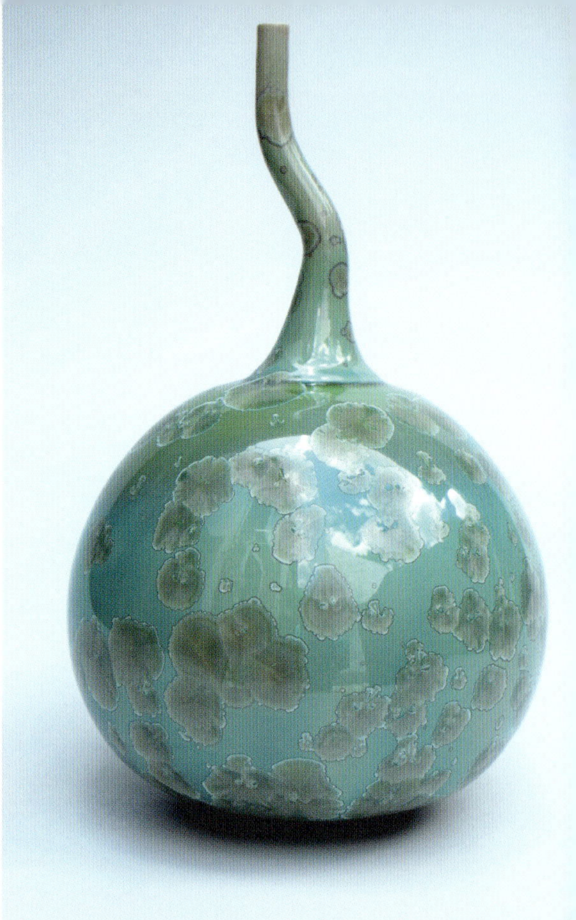

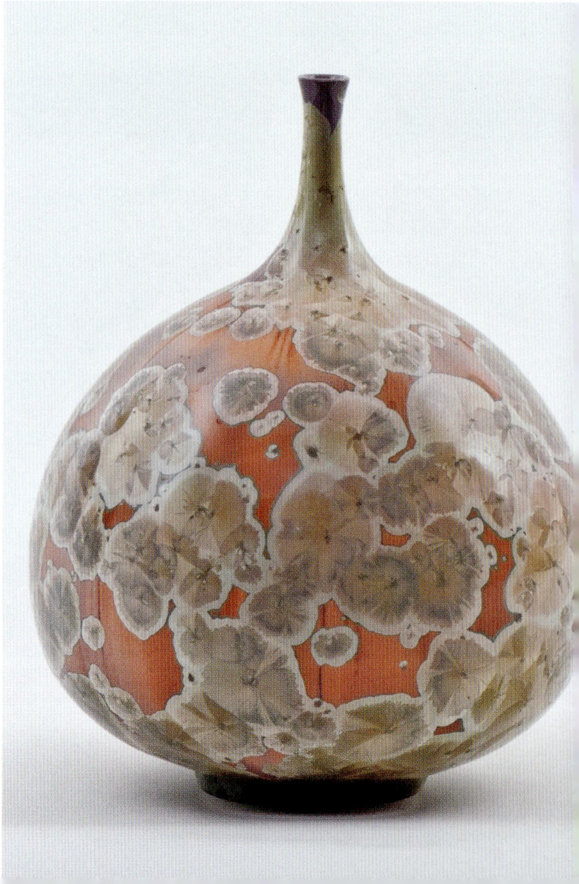

Introduction

LEFT Bottle with cobalt blue, oxidation fired.

TOP RIGHT Copper green bottle fired in oxidation.

BOTTOM RIGHT Bottle with rutile colourant, fired in reduction.

The first time I saw a crystalline glaze was mind-boggling. Although I had been working exclusively in porcelain, I had never seen anything quite so intriguing. Images of thistledown blowing in the wind and floating galaxies came to mind. How did the potter do it? I wanted to find out.

Since that first exposure at a craft show 40 years ago, I have seen many different types of crystalline glazes and met hundreds of potters working in this technique. And yet, I still find the process fascinating.

Creating crystalline glazes involves more than a methodical procedure comprising recipes and firing schedules. It requires an understanding of what crystals are, how they grow, how the ingredients in the glaze work together, and how each material reacts in the melt. Knowing the glazes well and mastering when to use them are skills that require research, time, and sensitivity to artistic expression. Perhaps working with porcelain and using crystallines is where technology and art are most compatible. It seems natural to pair the two.

However, using a crystalline glaze does not guarantee that the piece will be aesthetically pleasing. Too often, it camouflages an unsuitable form or a poorly made pot.

Some potters, although intrigued, have been too intimidated to try this technique. It has been described as difficult, with a high failure rate. This book attempts to demystify the process and provide an understanding of the materials. It offers information needed to work in this medium and minimise losses. I hope you find this guide useful for developing your own new and exciting glazes and making uniquely beautiful pots.

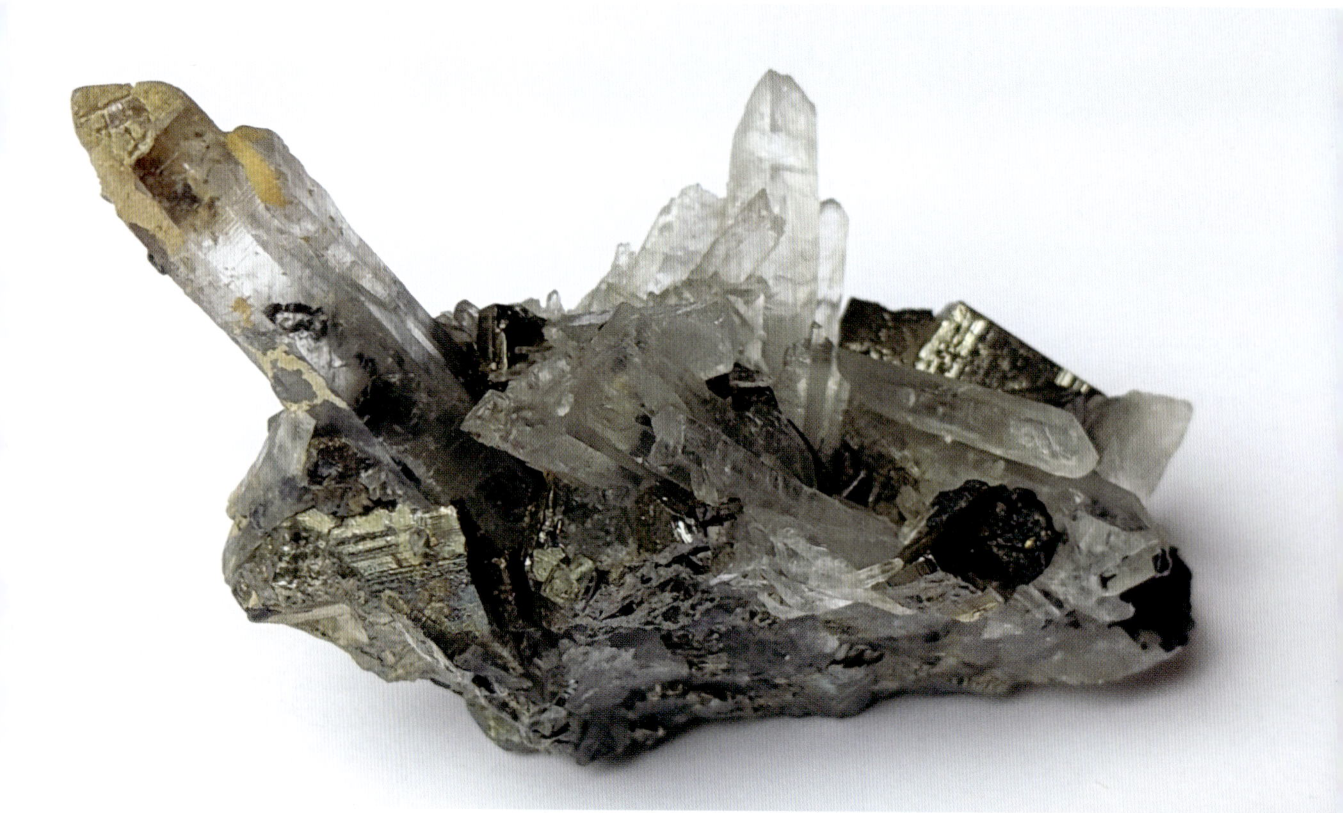

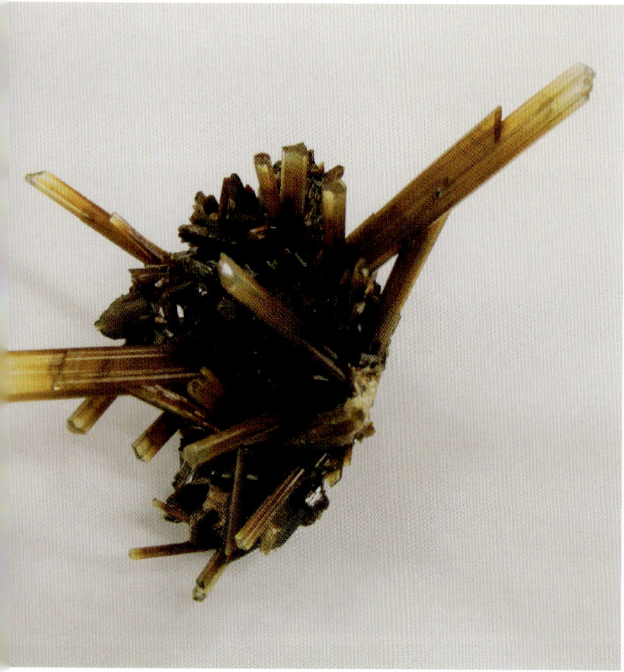

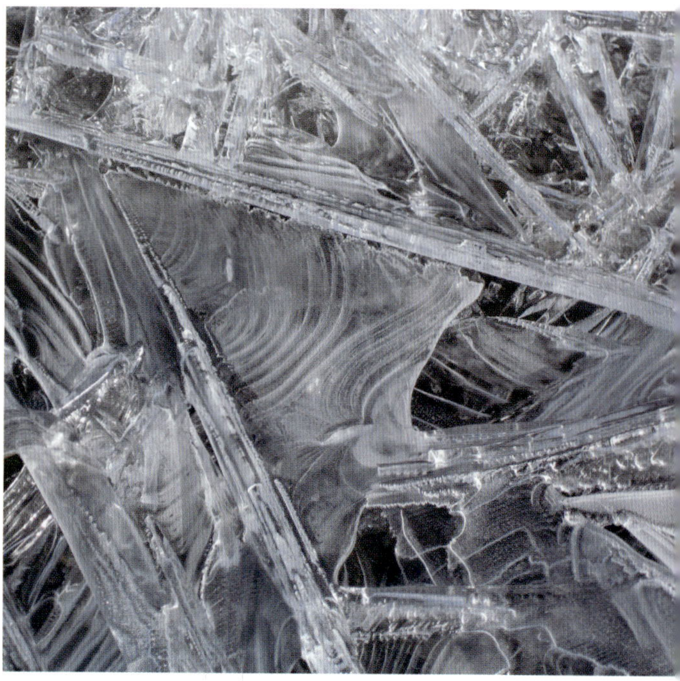

1

Natural crystals

TOP Quartz crystals, (white) with mica crystals (gold). Ontario, Canada.

BOTTOM LEFT Selenite gypsum, Chihuahua, Mexico.

BOTTOM RIGHT Ice crystals in a puddle.

From snowflakes to diamonds, humans have always been fascinated with crystals. Throughout history, crystals have played an important role in religious and healing ceremonies. Imbued with mystical powers, forecasting the future, or used as body adornment, they have been valued, traded, and crafted into many of the basic materials of our daily lives.

But what are crystals? Simply put, they are substances whose elements are arranged in three-dimensional patterns that fit together and repeat themselves, giving their characteristic shape.

Billions of years ago, the Earth was a mixture of hot gases and dust remaining from the formation of our sun. As the materials amalgamated and cooled, the Earth grew and distinct layers formed. The Earth's mantle consists of magma and super-heated rock with temperatures ranging from 1000°C near the crust to 3700°C closer to the core (1832°F to 6692°F). This differentiation occurred because of the varying densities of the different minerals.

Differences in density and pressure can sometimes cause magma to rise to the surface, where it erupts as volcanic lava. As the lava cools, the movement of its constituent molecules slows and, depending on its composition, it solidifies at specific temperatures and forms lattice structures we call mineral crystals. Additionally, rocks can be described as aggregates of minerals. All minerals are crystalline.

The crystal lattice takes six specific forms: cubic, hexagonal, tetragonal, orthorhombic, monoclinic, and triclinic. Each of these types contributes to the diversity of crystal shapes. The angles at which atoms are bonded and the distance between them create the geometric patterns characteristic of specific crystal types.

Most crystals require a medium in which to grow. Magma when cooled produces minerals. Water provides a growth medium for ice crystals and in water-based solutions, such as seawater, crystals form and grow through evaporation, becoming salt.

When solutions cool slowly and excess molecules can no longer be contained, they align themselves on a nucleation site, or 'seed' (a single crystal from which a larger one grows). Crystals cannot form within a liquid without some sort of seed.

When conditions are ideal, crystals may start growing in many different locations within a solution. They will continue to grow until something foreign gets in their way, or until the saturated material needed for growth is depleted. Many appear in clusters, and most rocks contain several different kinds of minerals.

Several factors, such as composition, temperature, pressure, and time, influence their formation, growth and type. Since most materials are more soluble in a liquid at higher temperatures, an increase in temperature leads to saturation. Larger crystals tend to grow if cooling takes place over a longer period of time, or if a solution is still saturated with growing material.

Smaller crystals develop when solutions cool quickly. For example, sugar derived from plants when melted and cooled rapidly forms a glass, and appears as transparent hard candy. The molecules do not have enough time to order themselves into a crystalline structure. If the same sugar is added to water in large enough proportions and cooled slowly, it will form a saturated or supersaturated solution and will, over time, grow the beautiful crystals aptly named 'rock candy'.

Some crystals are so small that a powerful microscope is needed to see them. Others are as huge as a boxcar. Regardless of size, each mineral produces crystals with the same molecular structure.

Crystal growth is also dependent on space. If the space is confined, they compete for room to develop and may form imperfect, irregular shapes. In contrast, open cavities, such as those found in geodes, allow for the growth of well-formed, often spectacular crystalline structures.

Almost all inorganic compounds are crystalline at room temperature, with the exception of glass and liquids. When glass cools, it appears to solidify, but whether it is a true solid is still a matter of scientific debate. It does not have a crystalline structure and its molecules are random.

Rocks that are deeper in the earth are subjected to heat and pressure and can react to form new minerals or recrystallize. Examples include microscopic calcite crystals in limestone growing larger, limestone metamorphosing into marble, or carbon becoming diamond under immense pressures.

Crystals play an integral role in our everyday lives, from the accurate splitting of seconds in advanced timekeeping instruments to the launching of a spacecraft. Radios, televisions, cell phones, fibre optics, computers, auto ignitions, hearing aids, pacemakers, and innumerable other devices incorporate solid-state components (silicon chips). Crystals help regulate power supplies to our cities and control our mass transportation systems. They are used in ultrasound equipment and lasers for diagnosing diseases. The list is endless.

There are just not enough natural crystals available to satisfy the needs of technology. They may be too valuable, the wrong size or shape, or not pure or perfect enough. More ore deposits must be found, explored, mined, and developed to satisfy the demands of advancing technology. Because of this, some crystals used by industry are grown in commercial laboratories using natural minerals and elements to suit specific needs. Examples include gemstones such as diamonds, sapphires, rubies, and emeralds.

Crystals have been forming on Earth for billions of years, and new crystals continue to form, both in nature and in laboratories. They shape the Earth's landscape by contributing both beauty and important materials to our daily lives.

TOP RIGHT Amethyst with hematite, Ontario, Canada.

MIDDLE RIGHT Brucite, Pakistan.

BOTTOM RIGHT Pyrite, Bolognesi, Peru.

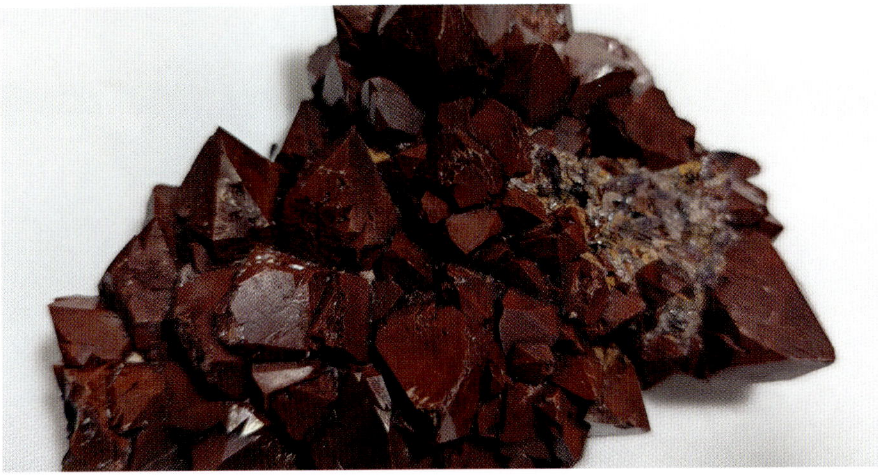

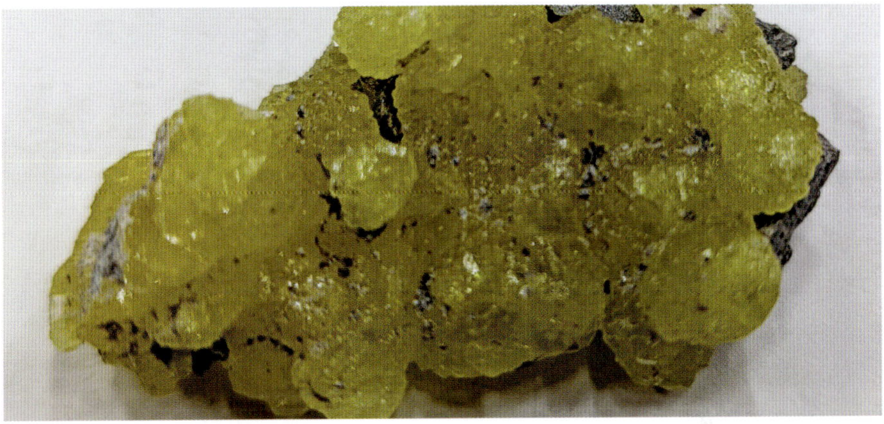

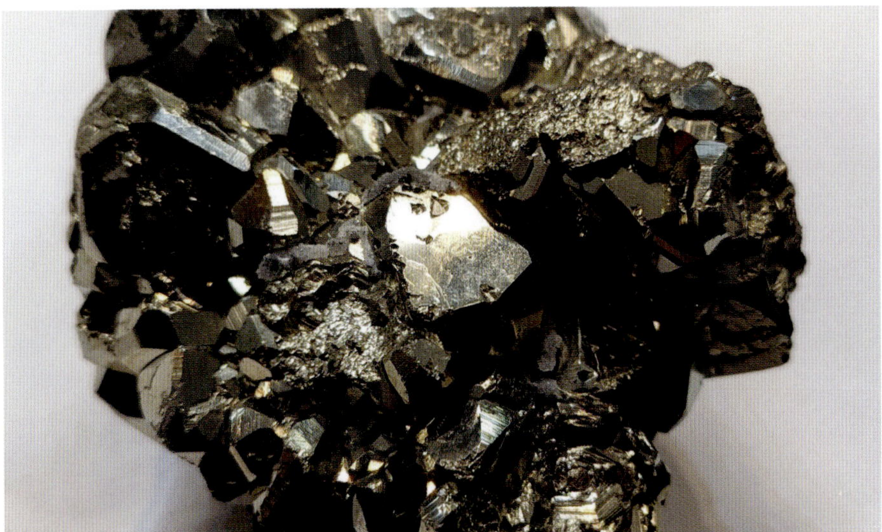

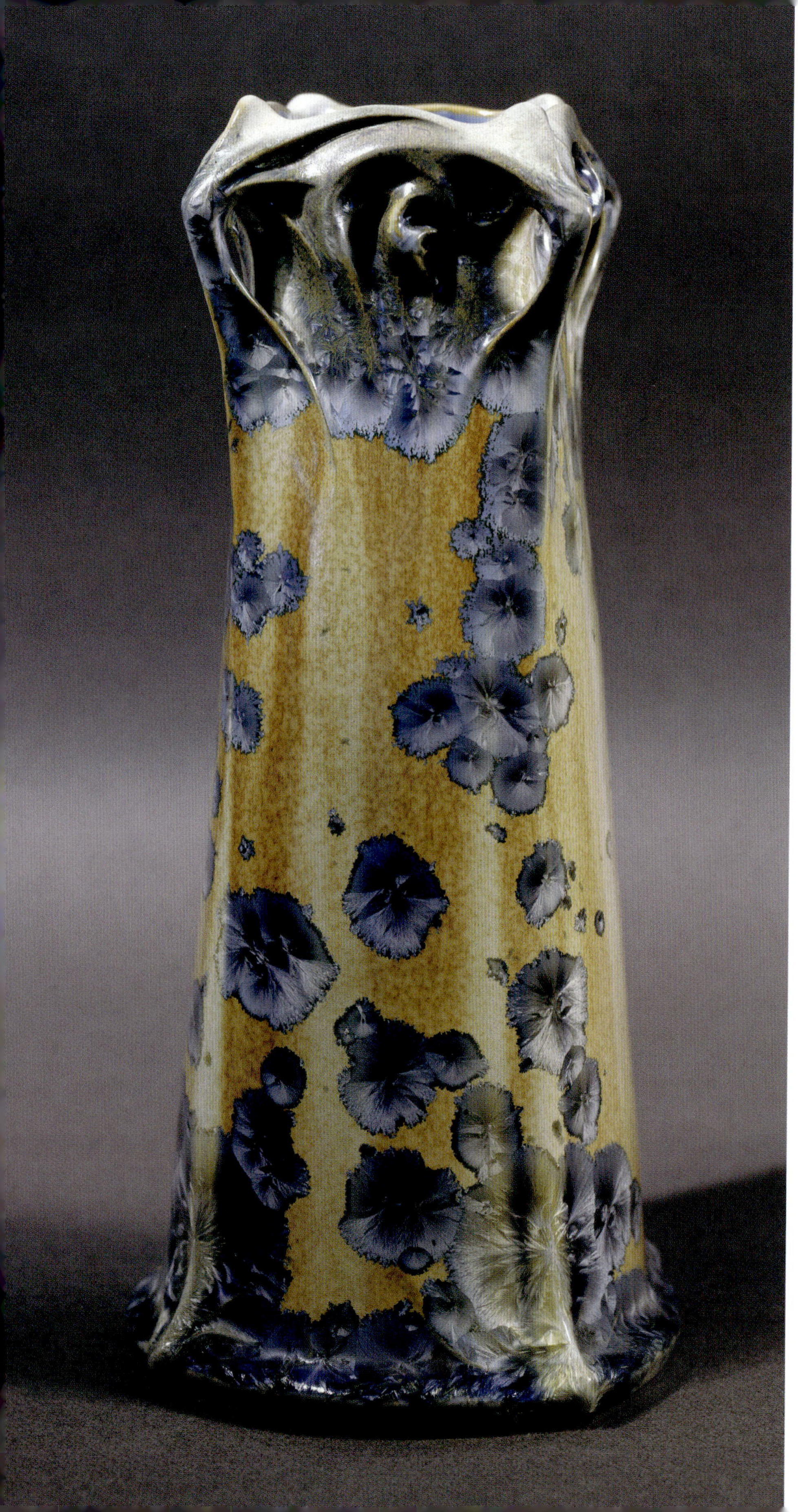

LEFT Art Nouveau style vase, 1904, by Hector Guimard for Sevres Manufacture, France, early twentieth century. Image courtesy of DeAgostini Picture Library/ Scala, Florence.

2

The history of crystalline glazes

Pottery making, one of the world's oldest crafts, predates history; comparatively, crystalline glazing is in its infancy. Until 140 years ago, there were no crystalline glazes or, if they existed, they were probably not understood, perhaps developing accidentally while the potter was trying to prolong the cooling of the kiln.

Most crystalline glazes are created on porcelain. China clay, known internationally as kaolin, is more commonly found in China than elsewhere, and the Chinese have been credited with the discovery of porcelain. With proper additions to the clay and higher temperatures in their kilns, Chinese potters were able to achieve hardness, whiteness, and translucency. As porcelain became more widely available, pottery became finer and the glazes suited to this form grew in demand.

The Song Dynasty (AD 960–1279) was known for its iron-rich *tenmoku* glazes, described as 'hare's fur', which contained microcrystals. The ancient Chinese oil-spot glazes also contained very small crystals, which were most likely unintentional.

Development of crystalline glazes in continental Europe

When kaolin deposits were discovered near Alençon and Limoges in France in around 1768, porcelain manufacturing in Europe started using this local material. Because kaolin is iron free and allows the clay to be fired to a higher temperature, the objects became more refined, and new glazes were developed.

In 1800, Alexandre Brongniart became the director of Sèvres National Porcelain Factory in France. During the five decades he worked there, he was the first to successfully oversee the manufacturing of European hard-fired porcelain; he also oversaw the creation of new enamel colours, and installed more efficient and cost-effective kilns capable of reaching higher temperatures.

Charles Lauth and George Dutailly, two chemists at Sèvres, discovered crystal deposits in glazes made with zinc silicate and titanium dioxide, but they chose not to develop the procedure. Instead, they issued a warning to other chemists about these 'contaminants' and how to avoid the 'problem'. A cup they made with a crystalline glaze was donated to the manufactory's museum. This cup and the date 1885, established the documented beginnings of crystalline glazes produced at Sèvres.[1] However, the factory did not start

[1] Axel Schmidt-Walguny, ed., *Kristallglasuren, Kreisverwaltung des Westerwaldkreises*, Montabaur, 1987, p.13.

formally producing crystalline glazes until 1897. The artist-potter Taxile Doat began working at Sèvres in 1877; although his work style was academic, he experimented with crystalline glazes. Doat left Sèvres in 1905 and emigrated to America where he continued working with these glazes at the University City Pottery in St Louis, Missouri.

Adolphe Clément, a chemist at the Royal Copenhagen Porcelain Manufactory, Denmark, disregarded the warning and began experimenting with zinc silicates. In 1886, he donated examples of his results to the city's mineralogical museum. Seven years after leaving Royal Copenhagen, he published a report of his research into these glazes to the Congress for Applied Chemistry in Vienna.[2]

A chemist named Valdemar Engelhardt succeeded Clément at the Copenhagen factory in 1891. He developed crystalline glazes for artistic application, which became internationally known. At the Chicago exhibition of 1893, they were much acclaimed.[3]

RIGHT Vase with twisted base, 1894-1900. Designed by Valdemar Engelhardt for Royal Copenhagen. Copenhagen, Denmark. Image courtesy of The Newark Museum of Art/Art Resource/Scala, Florence.

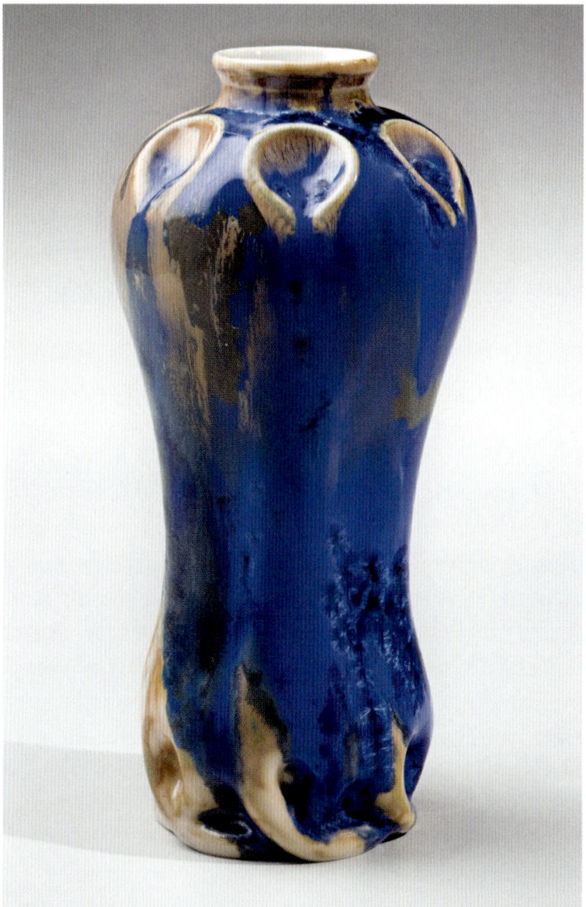

[2] Antoinette Fäy-Hallé, *Porcelain of the Ninteenth Century*, trans. Barbara Mundt, Rizzoli International Publications, New York, 1983, p.260.

[3] Marete Bodelsen, *The Royal Copenhagen Porcelain Manufactory 1775-1975*, The Manufactory, Copenhagen, 1975, p.66.

The Royal Porcelain Manufactory of Berlin was the first to market crystalline glazes commercially. In 1878, the company established the Techno-Chemical Research Institute. Albert Heineche, head of the Institute from 1888–1914, started concentrating on crystalline glazes in around 1898.[4] 10 years later, Wilhelm Pukall, a chemist at the Manufactory and head of the Royal Ceramic College, also worked with crystalline glazes and published the results.

The Meissen Porcelain Manufactory in Germany first began producing crystalline glazes in 1898 and exhibited them at the Paris Exhibition of 1900.[5] Also at the turn of the century, the Rörstrand Porcelain Manufactory in Sweden started producing Japanese-style porcelain forms with crystalline glazes.

From the mid-eighteenth century to the outbreak of World War I, most ceramic work was made in large factories and was usually cast. Separate workers performed each step along the road of production; an individual piece might be handled by dozens of workers. An artist designed the object, someone else prepared the clay, another cast the piece, a chemist made the glaze, and others were responsible for glazing and firing. Most of the ware produced around this time was very highly decorated with paintings or transfers of florals, animals, and a myriad of other images in the preferred style.

It was not until the Art Nouveau movement (1890–1910) that a simpler glaze became popular. The single glaze on a straightforward form began to appeal, perhaps as a relief from the overdecorated work preceding it. Crystalline glazes, with their subtle colour variations, random occurrence, and impressionistic appearance, better suited the natural and sensuous lines of the Art Nouveau style. Industrial ceramicists, interested in the feasibility of commercial production, began experimenting with these glazes.

Development of crystalline glazes in England

As information on crystalline glazing became more widely known, Pilkington's Tile and Pottery Company and the Royal Doulton Company in Great Britain began developing the technique.

In 1903, Pilkington created several crystalline lines, including 'Starry Crystal', so-called because of the jagged shape of the small crystals, and 'Opalescent Crystal', a milky opalescent with clusters of small matt crystals. Despite these innovations, Pilkington's became better known for their aventurine glazes – lower temperature matt glazes containing suspended glittering specks. Being more controllable, these glazes were more commercially viable.

[4] Fäy-Hallé, *Porcelain*, p.260.

[5] Hugh Wakefield, 'Modern Industrial Ceramics (from 1860)', in Robert J. Charleston, ed., *World Ceramics: An Illustrated History from Earliest Times*, Hamlyn, London, 1968, p.304.

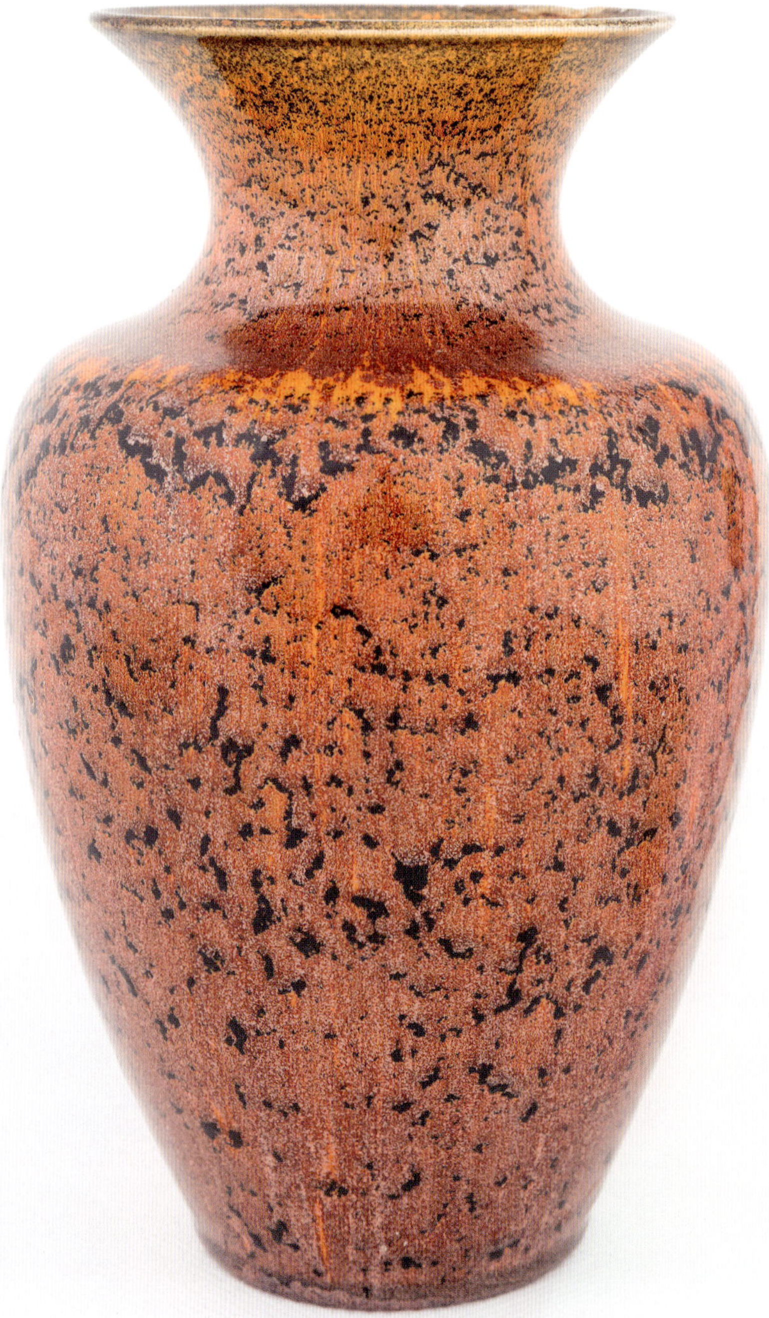

LEFT Crystalline vase by Pilkington's Tile and Pottery Company. Image courtesy of Salford Museum & Art Gallery, Salford, England.

Royal Doulton was soon internationally recognised for outstanding achievements with crystalline glazes. At the Brussels Exhibition in 1910, every piece was sold within hours of the opening. Using a high-temperature porcelain body fired to 1400°C (2550°F), they created a glaze that featured larger single crystals floating in a glassy background. With the outbreak of World War I, Royal Doulton stopped producing these expensive and unreliable glazes.

Ruskin Pottery in West Smethwick, near Birmingham, was one of the last British art potteries to experiment with crystalline glazes. In the mid 1920s they started producing a variety of unusual glazes, including micro, macro, aventurines, and lustres. This was at a time when most potteries had stopped producing these glazes because of their expense and unpredictability. When Ruskin closed in 1935, the formulae for the glazes and all the pottery documentation were deliberately destroyed to prevent replication of the unique Ruskin products.

RIGHT Vase, 1926. Ruskin Pottery (1901-1935). Image courtesy of Museum Associates/LACMA/ Art Resource NY/Scala, Florence.

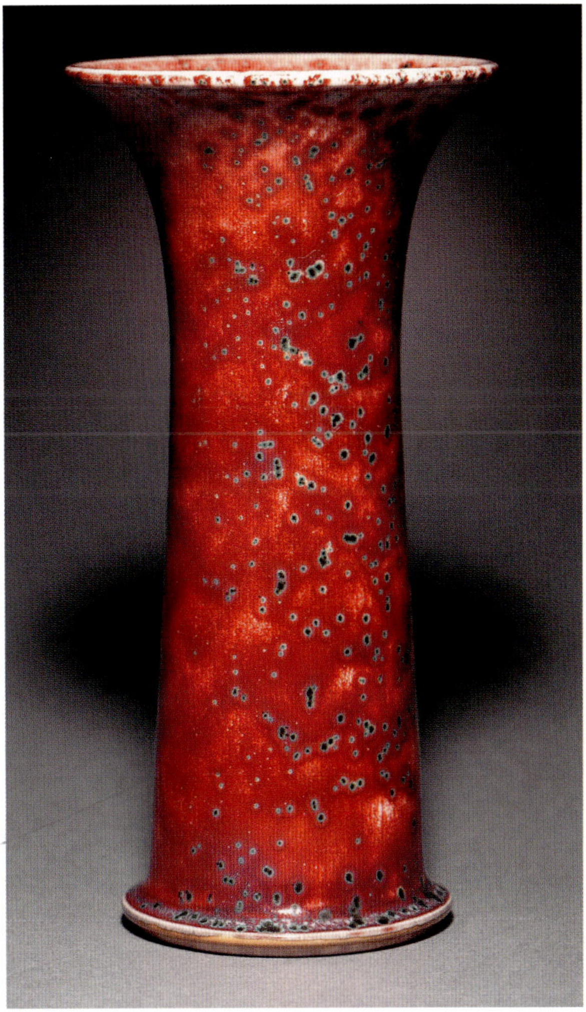

Development of crystalline glazes in the United States

In 1880, Maria Longworth Storer founded Rookwood Pottery in Cincinnati, Ohio. It was the first large manufacturing enterprise founded and owned by a woman in the United States. Within a decade, the company dominated the American art market and gained international acclaim, rivalling European and Asian firms.

Rookwood achieved an international reputation and won a gold medal at the Paris Exhibition of 1889 with a crystalline glaze called 'Tiger Eye',[6] which produced a golden shimmer from deep within the glaze. This and the company's other popular glaze, 'Goldstone', were really aventurines. Rookwood hit its peak just before World War I. After the war, only the Tiger Eye glaze was revived. The aventurines were difficult to produce consistently, and Rookwood stopped making them in 1934. The company, however, remains in operation to the present day.

The Fulper Pottery Company of Flemington, New Jersey, may have been established as early as 1805. Initially, it made utilitarian ware such as crocks, filters, and cooking utensils. In 1909, it entered the art pottery field with a more diversified line called VaseKraft, which was decorated exclusively in ornamental gloss, matt, monochrome, lustre, and crystalline glazes.[7]

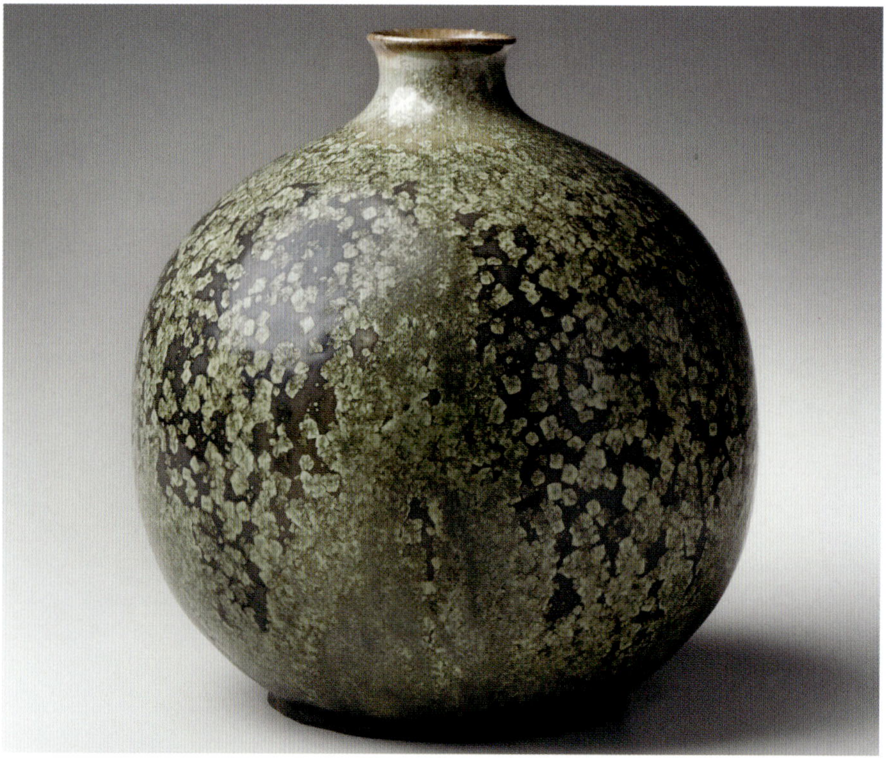

LEFT Globular vase with 'leopardskin crystal' glaze, 1914. Manufactured by Fulper Pottery Company. Flemington, New Jersey. Image courtesy of The Newark Museum of Art/Art Resource/Scala, Florence.

RIGHT Vase, c.1912-1915. Fulper Pottery Company (1899–1935). © The Metropolitan Museum of Art/Art Resource/Scala, Florence.

[6] Wakefield, 'Modern Industrial Ceramics', p.313.

[7] Paul Evans, *Art Pottery of the United States: An Encyclopaedia of Producers and Their Marks*, Scribner, New York, 1974.

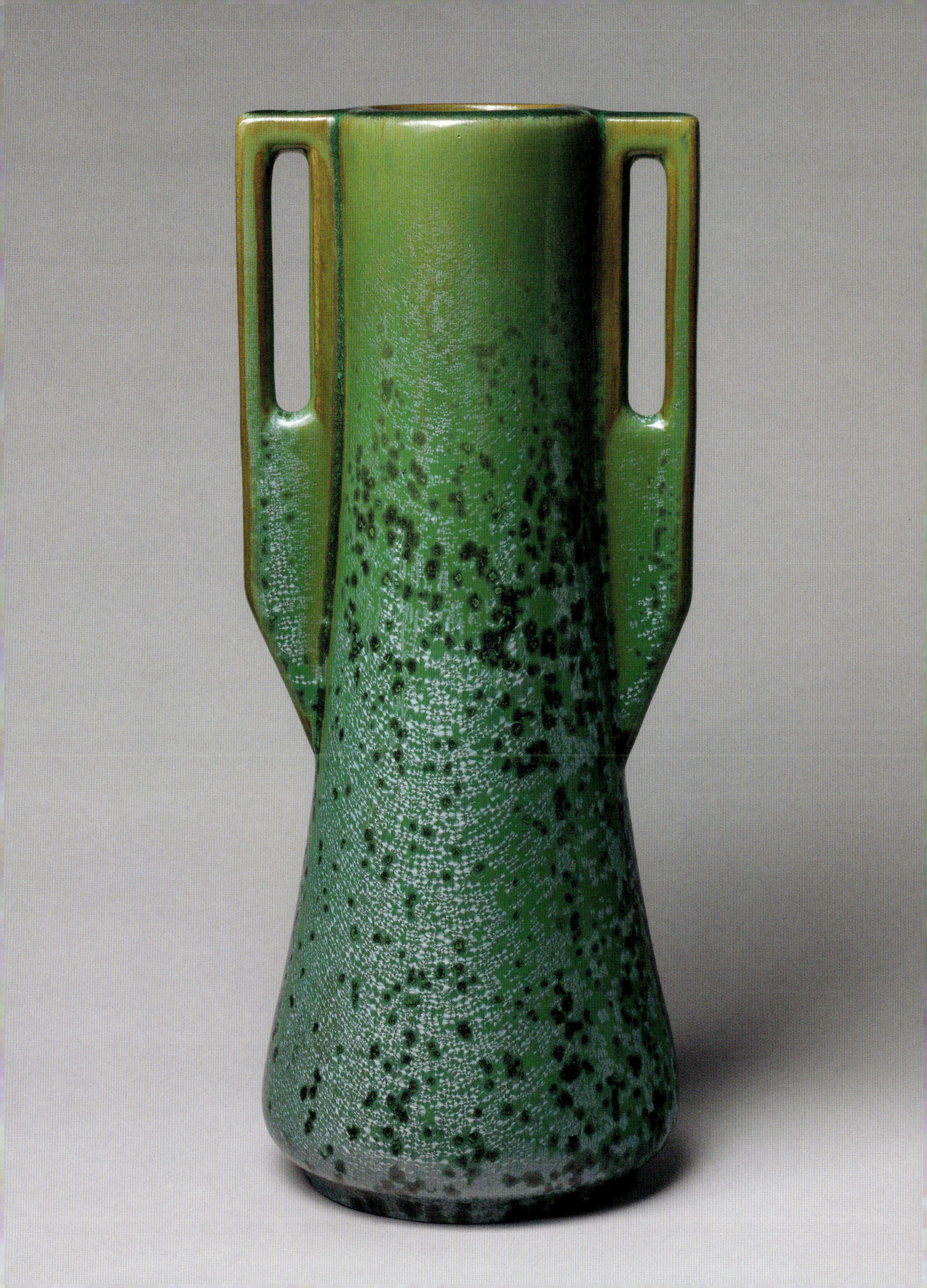

Teco Pottery in Illinois was originally the art pottery line of the American Terra Cotta Tile and Ceramic Company, producers of sewer pipe, brick, and terracotta. At the turn of the twentieth century, the company developed Teco Green, a matt glaze used on organic shapes. Successful experiments led them to produce aventurines and experiment with crystallines. By 1901, the company had created minute iridescent crystals and became the first art pottery in North America to duplicate the large crystals developed earlier in Europe. No two pieces were alike. Teco Crystal, as the ware was called, was marketed nationally. Producing this kind of ware on a commercial basis, however, was just too expensive, so these designs were abandoned. Once again, the single Teco Green glaze dominated the line.[8]

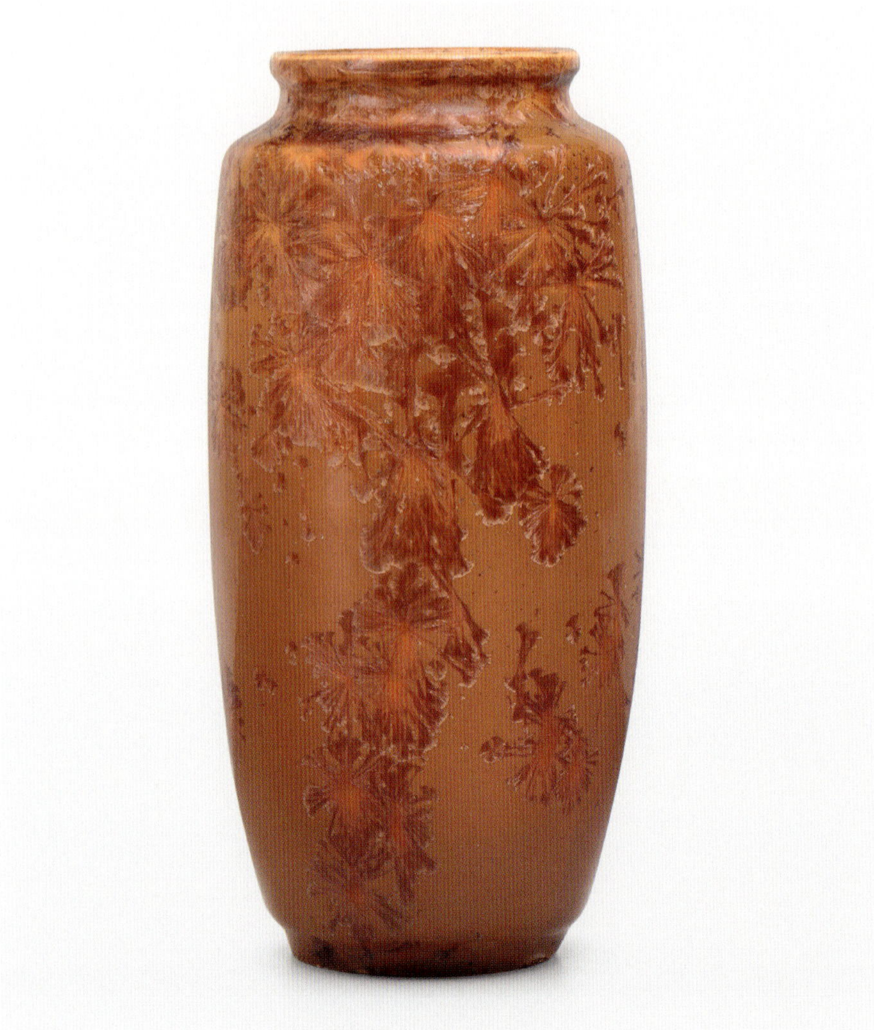

LEFT Teco Crystal vase *c.*1904. Collection of Richard D. Mohr. © Toomey & Co. Auctioneers (Rago/Wright).

[8] *Ibid.*, p.280.

Competition among the European porcelain factories was so keen that chemists kept most of their recipes and methods of working a secret. As a result, American companies were forced to conduct their own research and experiments. The American Ceramic Society, formed in 1899, comprised ceramicists working for industry, universities, and the government. Discoveries were published in the *Journal of the American Ceramic Society*. In 1904, several articles on crystalline glazes were circulated. Recipes were included and potters were encouraged to experiment with these new glazes. F. H. Norton, a frequent contributor to the magazine, wrote several articles on the subject. In 1937, he presented a paper to the society entitled, 'The Control of Crystalline Glazes'. Norton's experiments used seeding to produce crystals at desired locations and he was also able to control the size and shape of crystals by varying time and temperature.

After the turn of the twentieth century, most of the porcelain manufacturing factories in both Europe and America were experiencing financial problems with crystalline glazes. Production costs for individually glazed pieces were relatively high, and the technique was unpredictable. Mass-produced objects replaced elaborately moulded designs, and the complex glazes were replaced with simpler ones that did not require such careful processing. With the outbreak of World War I, most companies had stopped making these glazes.

The idea of the artist-potter changed dramatically during the 1920s and 1930s. A worker being responsible for a single aspect of the assembly lost its appeal. Individual potters now performed nearly all the production processes. They dug their own clay, designed and made the pots, then glazed and fired them in their own personally constructed kilns. The factories could mass-produce the functional ware, but the artist-craftsperson rose in defiance of the materialism of industry and its insensibility to beauty. Adelaide Alsop Robineau (1865–1929) was one of the first to create her own objects on the potter's wheel, decorate and glaze them with her own glazes, and fire them in her kerosene kiln.

Adelaide Alsop Robineau

Artistically talented, Adelaide Alsop learned from books how to decorate porcelain, and eventually became proficient enough to teach the craft. In 1899, she married Samuel Robineau, a Parisian who shared her enthusiasm for fine porcelain. Around the turn of the twentieth century, the couple established an Arts and Crafts magazine, *Keramic Studio*, from their home in Syracuse, New York. During the thirty years of her editorial leadership, the success of the publication could be attributed to Adelaide's many features, editorial comments, and 'how to' articles. In her research, she came across pieces written by Taxile Doat, the eminent ceramicist from Sèvres National Porcelain Factory. The Robineaus contacted Doat in 1903, and Samuel translated a series of his articles for the magazine.[9]

[9] F. H. Read, 'The University City Venture' in *Adelaide Alsop Robineau: Glory in Porcelain*, ed. Peg Weiss, Syracuse University Press, Syracuse, New York, 1981, p.95.

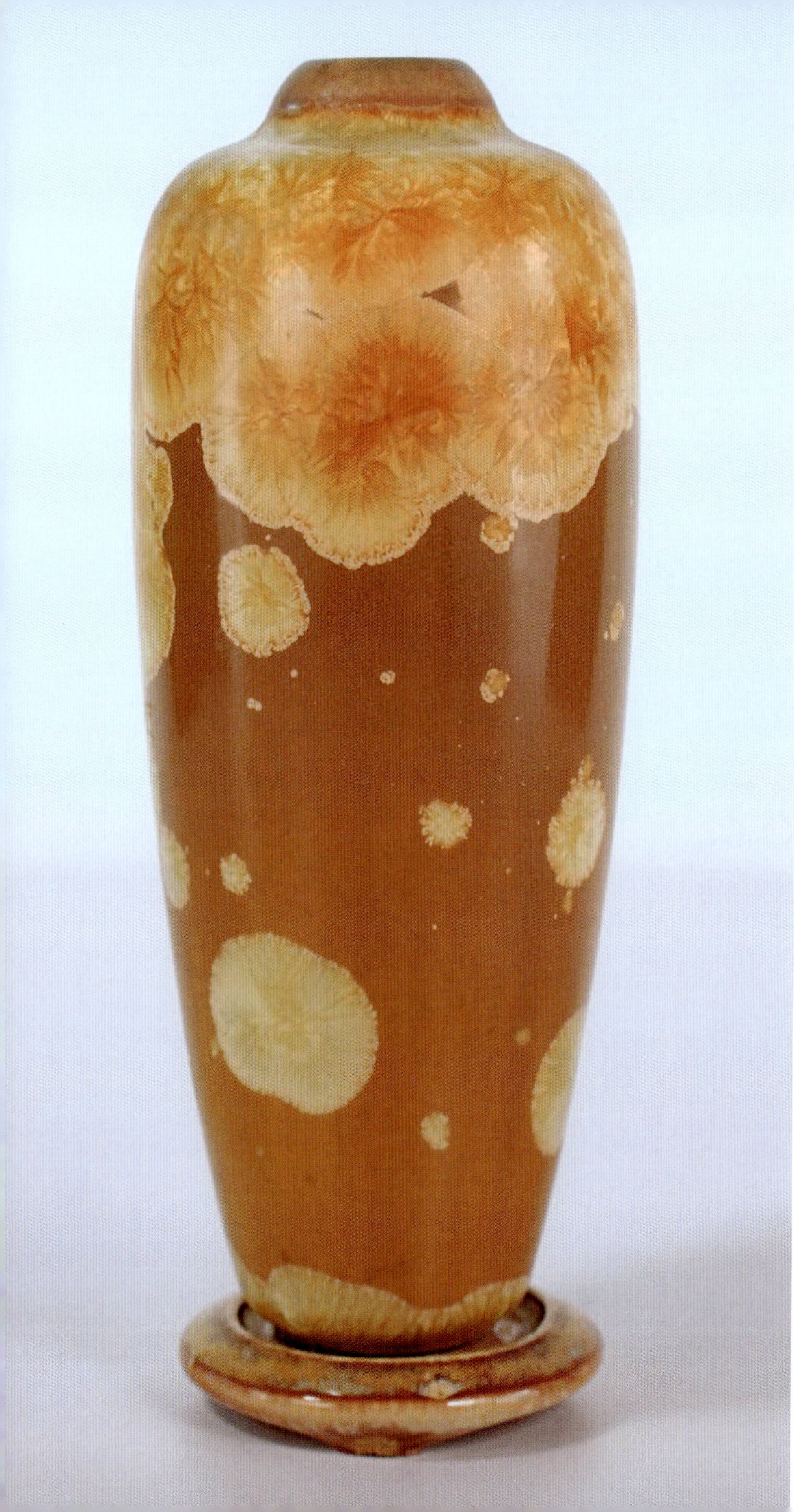

LEFT Adelaide Alsop Robineau. *Vase with stand*, 1910. Porcelain. Image courtesy of Everson Museum of Art. Museum purchase, 16.4.15.

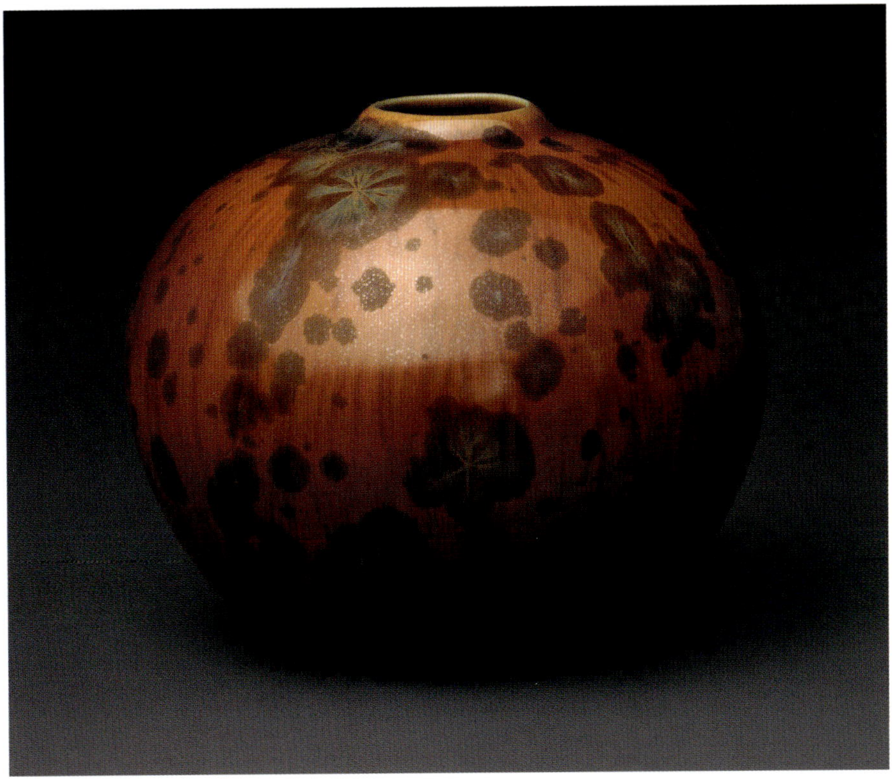

RIGHT Adelaide Alsop Robineau. Vase, 1905. © The Metropolitan Museum of Art/Art Resource/Scala, Florence.

Fascinated by high-fired porcelain, Adelaide enrolled in a summer course at the pottery school at Alfred, New York, taught by Charles Binns. Her husband helped her set up a kiln and became her technician. Adelaide experimented with different clay bodies and developed an unusual range of both fixed and flowing glazes. She also started experimenting with crystalline and matt glazes. In volume 3 of *Keramic Studio*,[10] she published an article on the Royal Copenhagen porcelains, which included crystalline glazes. She began to work mainly in porcelain herself, and made her first crystalline-glazed piece in 1904.[11]

In 1907, amateur potter and magazine publisher Edward G. Lewis discovered a vein of kaolin during the excavation of his publishing headquarters at St Louis, Missouri. He decided to build a pottery on the site, focusing exclusively on porcelain. To launch the University City Pottery, as the enterprise was called, he invited Taxile Doat to head the School of Ceramic Art, also on the premises.

Lewis also invited Adelaide and Samuel Robineau to join the venture. This pottery, operating from 1909 until 1914, was an opportunity for the Robineaus to work with a great master in an ideal setting, with all materials supplied, and an annual income. However, financial difficulties plagued the University City venture. After a year, the

[10] Adelaide Alsop Robineau, 'Royal Copenhagen Porcelain', in *Keramic Studio*, vol. 3, May 1901-April 1902, pp.10-11. Accessed at library.si.edu/digital-library/book/keramic-studio

[11] M. Eidelberg, 'Robineau's Early Designs' in *Glory in Porcelain*, p.77.

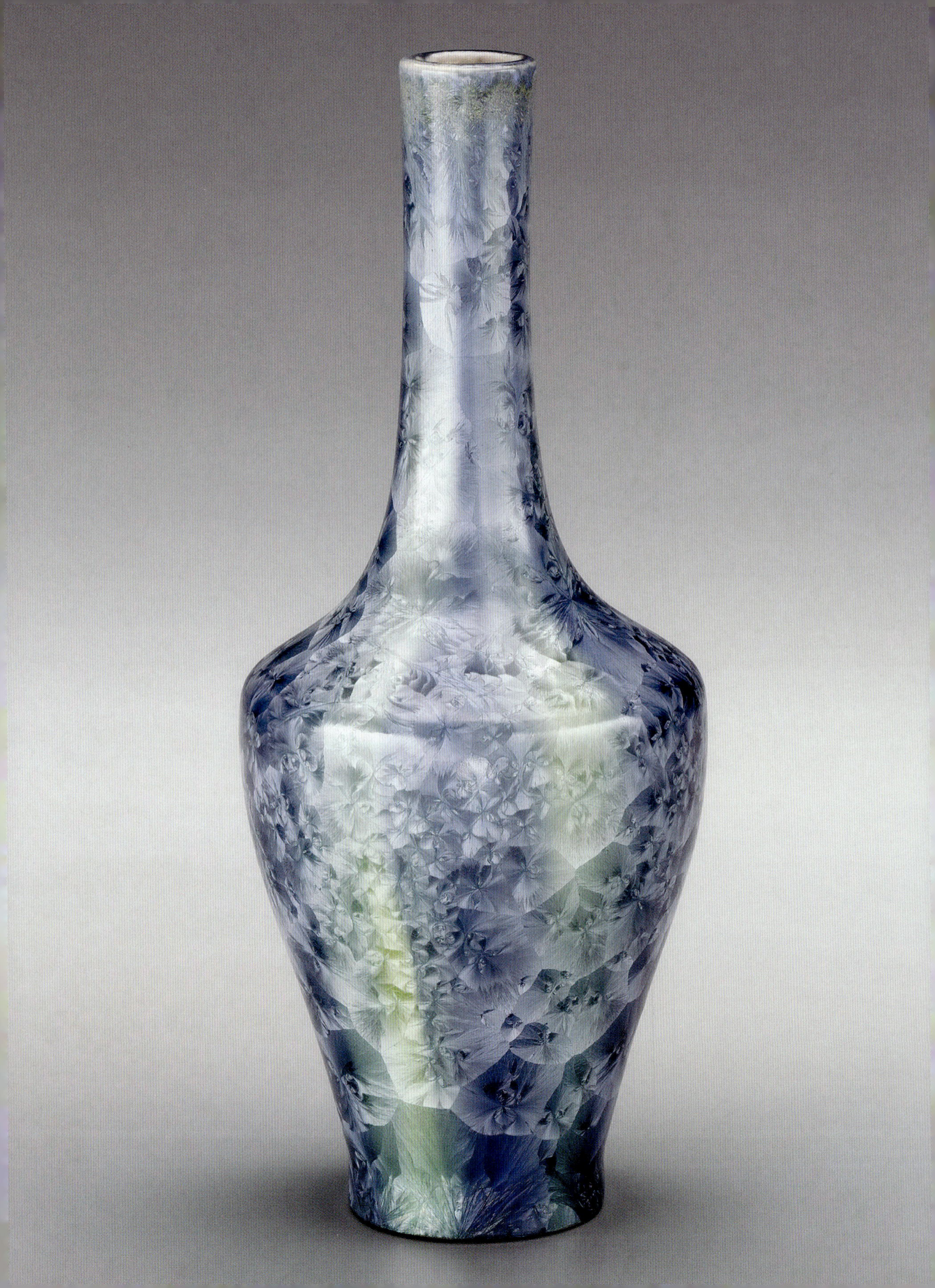

LEFT Vase, 1912-1913. © The Metropolitan Museum of Art/Art Resource/Scala, Florence.

Robineaus returned home to Syracuse, where Adelaide continued making pottery and taught at Syracuse University until her death in 1929. She was instrumental in developing a wide variety of beautiful and interesting glazes, many of them crystalline. Her experiments covered zinc silicate, flowing matt, and molybdenum. She even experimented with reduction-fired crystal glazes, and her variety and colour were unequalled for the time.

Esteemed ceramicists like Taxile Doat, William Grueby, and Charles Fergus Binns considered her among the avant-garde who pursued the most elusive and demanding techniques in ceramic history.[12]

Crystalline glazing in the twentieth century

By the turn of the twentieth century, crystalline glazes were gaining popularity in both Europe and North America. These new glazes better suited simpler shapes such as cylinders, bottles, gourds or ovals. With the rising of the Arts and Crafts movement, form without elaborate decoration became fashionable.

Perhaps the beginning of this aesthetic shift – at least in Great Britain – was thanks to Bernard Leach, who is widely regarded as the father of British studio pottery. Leach spent many years in Japan, where he became interested in Japanese pottery. On his return to England from Japan in 1920, Leach was accompanied by Shōji Hamada, an accomplished Japanese potter. Together they established a pottery at St Ives in Cornwall. Using natural, local materials for their clays and glazes, they allowed impurities in these materials to give character to the work.[13] Several of Leach's students and assistants went on to establish their own potteries. Leach's celebration of the individual craftsperson influenced other potters in Britain, continental Europe, and North America.

As the Arts and Crafts movement grew in the United States, clay became an acceptable medium for making art. Alfred University established its School of Clay-working and Ceramics in 1900; with pottery making now being taught at the university, the craft gained credibility. No longer were potters trained only to go into industry as designers. It was now acceptable for individual artisans to participate in all aspects of the work.[14]

The School's founder, Charles Fergus Binns, had a profound influence on the way his many students approached their craft. Binns stressed the importance of the individual potter's responsibility for all steps in the process. Binns believed that potters should mix their own glazes to enhance the individuality of the pot. His experience with this approach led to a greater public interest in crystalline glazes.[15]

[12] Thomas Piché Jr., and Julia A Monti, eds., *Only an Artist: Adelaide Aslop Robineau, American Studio Potter*, Syracuse University Press, Syracuse, New York, 2006, p.1.
[13] Wakefield, 'Modern Industrial Ceramics', p.314.
[14] Barbara Perry, 'American Ceramics 1920-1950', in *American Ceramics: The Collection of the Everson Museum of Art*, ed. Barbara Perry, Rizzoli International, New York, 1989, p.121.
[15] *Ibid.*, p.121.

The first PhD in ceramics in the United States was awarded to Herbert Sanders. He began researching crystalline glazes while at Ohio State University and developed his first recipes in 1931. From the late 1930s to the early 1970s, he wrote many papers and books – some about crystalline glazes, for which he later became known. He went on to become an important educator and researcher, teaching at Ohio State, Alfred University in New York, and California State University at San José. His book, *Glazes for Special Effects*, contains several chapters on crystalline glazing.[16]

Marc Hansen was a student of Herbert Sanders at San José. Hansen formulated his own frits, (a fluxing material), using them as all or part of the glaze. He maintained an extremely high standard for his work and developed some particularly beautiful crystallines. He went on to teach at Western Michigan University.

David Snair was a former student of Marc Hansen. In 1975, he wrote an article that appeared in the pottery magazine, *Ceramics Monthly* (USA).[17] In the article, Snair offered simple recipes using commercial frits and explained how to make collector dishes which would protect kiln shelves from the glaze run off. He also provided a detailed account of how to do a crystalline firing, opening the door for many potters tempted to take on the challenge.

It was Snair's article that first inspired me to try these glazes. With the help of an Ontario Arts Council Grant, I spent a year researching, experimenting, and contacting over a hundred crystalline potters to learn their work methods. I realised I had enough material to write a book. In 1997, *Crystalline Glazes* was published by A&C Black Visual Arts – the first complete book written on the subject. An updated second edition was published in 2005, which was later published by Herbert Press in 2024.

RIGHT *Small Porcelain Vase*, n.d., David Snair (d. 1999). Porcelain, 3 1/4 x 1 1/8 x 1 1/8 in. Gift of June Shealor, 86.2 © Canton Museum of Art.

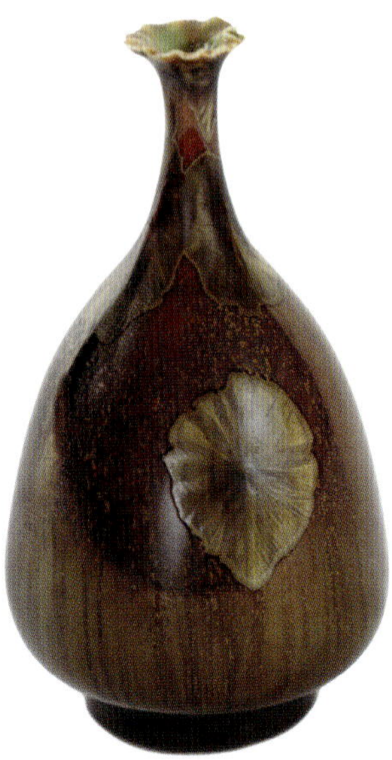

[16] *Ibid.*, p.177.

[17] David Snair, 'Making and Firing Crystalline Glazes', in *Ceramics Monthly* 23/10, 1975, pp.21-26.

Studio potters

Studio potters have been working with crystalline glazes for more than 50 years. Workshops and symposiums have been offered in this technique and some potters now use crystalline glazes exclusively. Ceramic magazines such as *Ceramic Review* (England), *Ceramics Monthly* (USA), and *Ceramica* (Spain) have published articles on this type of finish. Many new discoveries and experiments have taken place, allowing this process to become more widely recognised and appreciated. Beginning in the early 1980s, workshops and exhibitions have proliferated.

- The International Symposium of Crystalline Artists, founded in 1982, consisted of 20 potters who exchanged data on these glazes in a newsletter; the group disbanded in 1988.
- In 1987, a show of crystalline-glazed pottery produced by potters from Belgium, East and West Germany, France, the Netherlands, Austria, and Switzerland took place in Höhr-Grenzhausen, in what was then West Germany. The show received international attention. A catalogue was published featuring the artists and their work.
- In 1993, Derek Clarkson, a British potter, organised a show at the Manchester Art Gallery, England. Historic crystalline pieces were featured, together with the contemporary works of crystalline potters from England, Europe, the United States, Canada, and Japan.
- In April 2005, the International Exhibition of Crystalline Glazes took place in Vallauris, France, the 'city of potters'. Twenty-four potters from across Europe, Canada, and the United States participated in an exhibition. I was the guest speaker at the symposium.
- In September 2005, in Kansas City, Missouri, a two-day crystalline glaze conference and exhibition took place. More than 100 potters gathered for discussions and to exchange ideas. I also spoke at this event.
- In May 2006, a five-day workshop took place at the studio of Kris Friedrich in Palm Springs, California. Potters shared glaze recipes and brought their bisque-fired pots to be crystalline-fired in one of Friedrich's several kilns.
- Also in 2006, a show in Leeuwarden, Netherlands, featured the work of 14 crystalline artists.
- In March 2007, as part of the Canadian Clay Symposium, I conducted a five-day workshop on crystalline glazes at the Shadbolt Centre for the Arts in Vancouver, British Columbia.
- Bill Campbell of Cambridge Springs, Pennsylvania, held a workshop and exhibition at his studio and gallery in September 2007. Thirty-nine potters specializing in crystalline glazes participated.

RIGHT © Jose MARISCAL Paneque.

- In 2008, 18 potters from around the world exhibited their crystalline glazes at the Keramikmuseum Mettlach in Germany, which was sponsored by Villeroy & Boch as part of the 260th anniversary celebration of the company. The exhibition then travelled to two other German museums (Potzellanikan – the State Museum for Porcelain in Selb, and Museum Schloss in Fürsternberg) before making its way to the Dehullu Gallery in Gees, Netherlands in 2010.
- In 2009, 51 crystalline glaze potters participated in a three-day workshop at the studio of Phil Hamling in Warwick, New York. Demonstrations were given for several methods of achieving reduction of crystals in an electric kiln.
- In 2011, the National Council on Education for the Ceramic Arts (NCECA) took place in Tampa, Florida. As part of the conference, an exhibition of crystalline-glazed pottery was held at the nearby Dunedin Fine Art Center. Glenn Woods organised a panel discussion with renowned crystalline glaze artists Ginny Conrow, Phil Hamling, Jamie Kozlowski, Marsha Silverman, and me. We collectively addressed the various approaches crystalline glaze artists take and what they strive for as they seek perfect crystalline glaze results.
- In 2013, a five-day Congress of Crystalline Artists, organised by José and Maite Mariscal took place in La Bisbal d'Empordà, Spain with participants from around the world. The event featured an exhibition, presentations, demonstrations, a throwing contest, and a 'Night of Fire' celebration held in the town square.
- Almost every year at the Dunedin Fine Arts Centre, the Pottery Boys – Glenn Woods and Keith Herbrand – organise workshops, several of which feature crystalline glazes.

Crystalline workshops and exhibitions are popping up everywhere and now have become too numerous to mention. With recent advancements in computer technology, interest in crystalline glazes has increased, encouraging more potters to experiment with these fascinating glazes.

There have also been tremendous advances in kiln design and production. New insulation qualities and construction techniques now allow programmable electric and gas kilns to fire in oxidation and reduction and go through numerous ramps and holds (a ramp is a heat rise or fall, and a hold is a soak over an interval of time), all done automatically.

Crystalline potters today are finding unique ways of working and developing their own glazes and firing procedures, the methods being as varied as the crystals themselves. Consequently, the technology has advanced more in the last few years than in the past century. These glazes have become more popular than ever, with a greater number of potters now working with this amazing technique.

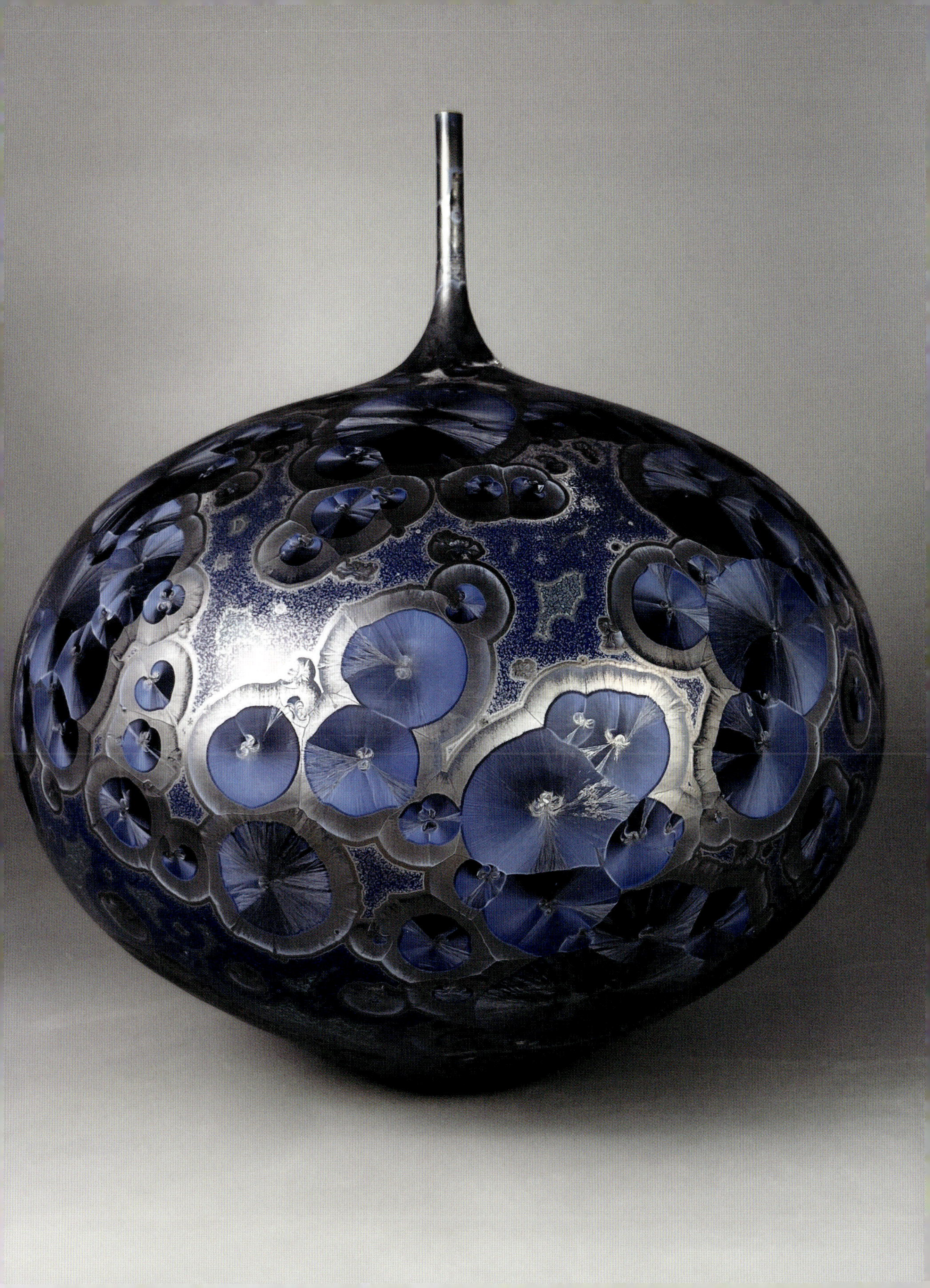

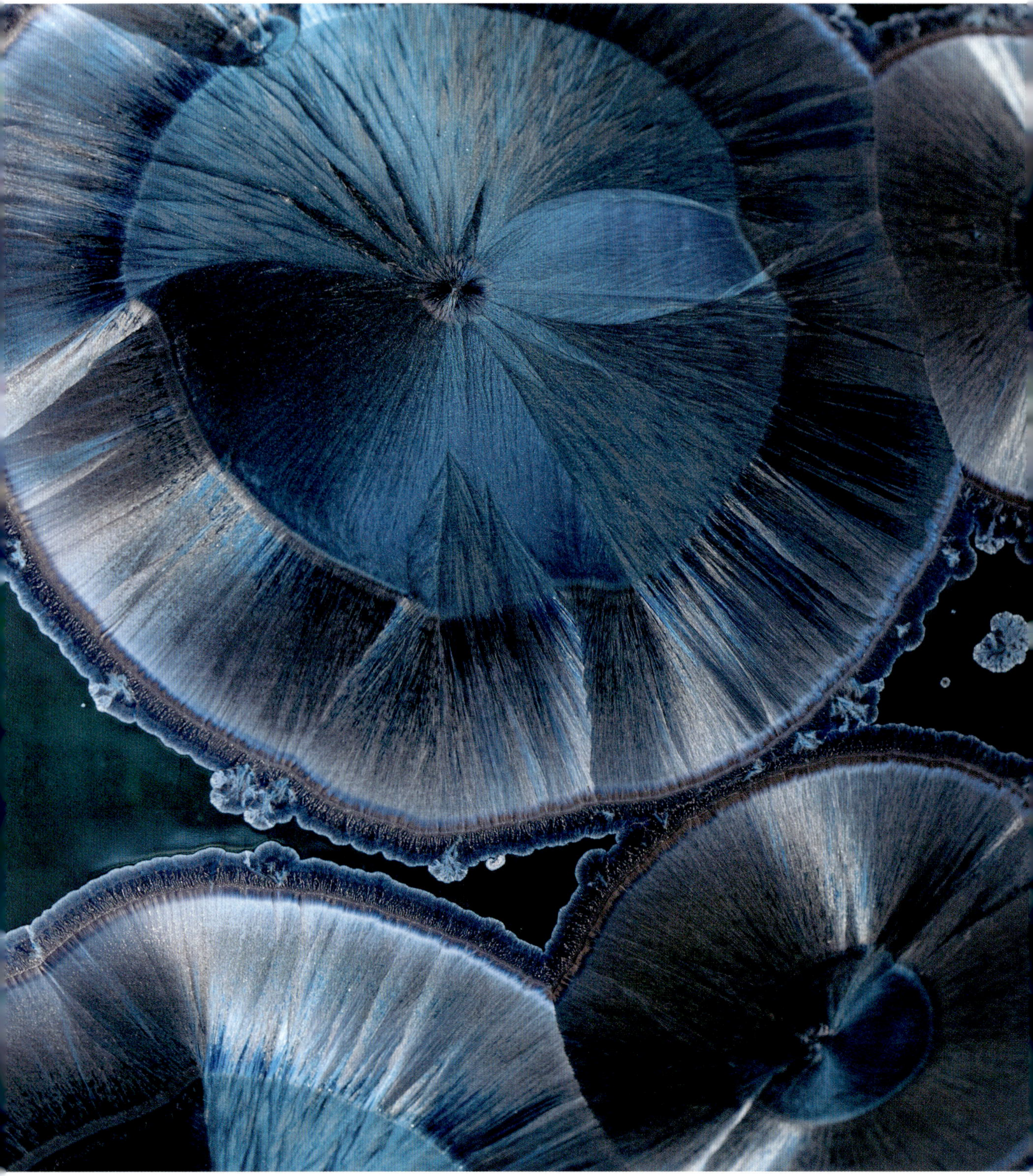

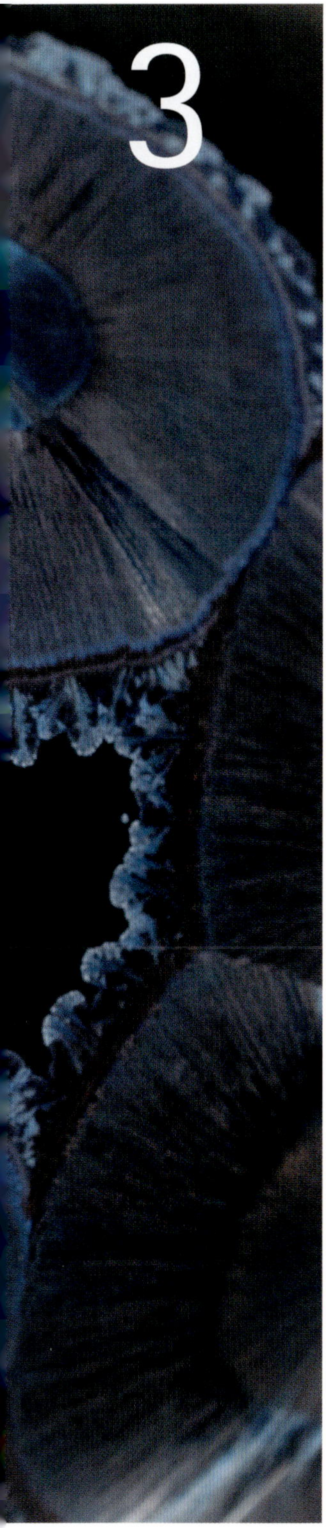

3

Growing crystals in a glaze

Crystalline glazes are an inspiration to potters who want to capture the essence of natural gems. Using appropriate recipes and a kiln to control heating and cooling, the results can include:

- matts – microscopic crystals that can't be distinguished by the naked eye;
- microcrystalline – distinguishable tiny specks that cover the whole surface of a pot;
- macrocrystalline – spectacular individual shapes that are clearly visible.

A glaze that contains little or no alumina fires clear and glossy and tends to run; it feels and looks like glass. When alumina is present, viscosity increases, helping the glaze adhere to the pot. The more alumina, the more matt and dry the glaze becomes. Crystalline glazes contain very little alumina because it is difficult for crystals to form in a viscose liquid. The glaze must be fluid at higher temperatures for them to separate, align, and then grow.

Recipes are made up of zinc oxide, silica, and an assortment of fluxes to increase the melt. Several other minerals are added to give colour and shape. When heated to a liquid, the molecules of the various minerals randomize and move freely.

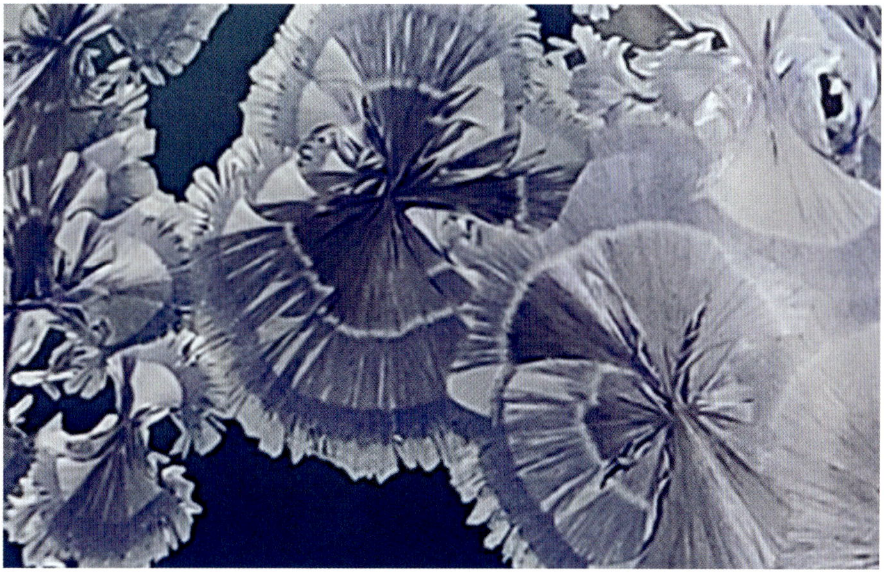

As the temperature drops to the crystal-growing range, the zinc silicate ignites, producing thin, needle-like crystals. As the temperature continues to drop, the crystals come together and nucleate to a seed consisting of an undissolved material. As the temperature cools further, the glaze starts to gel, and more time is needed for more complex structures. As cooling continues, the glaze eventually freezes. Every glaze has its own freezing point, but the freezing point for zinc silicate crystals is about the same.

It is important to know the maximum and minimum temperature limits of a crystalline glaze. At maximum temperature in the growth cycle, crystals are just beginning to form. If rapid cooling takes place, few or very small crystals will develop. A slower cooling period produces larger crystals. Once the growth limits have been determined, the temperature is either held constant or allowed to fluctuate within the set range for a predetermined amount of time.

With standard glazes, the melting point is the approximate maturing temperature of the glaze and represents the uppermost limit to which the glaze is fired. In contrast, a crystalline glaze is fired higher than the melting temperature to liquefy it completely. This could be at least 222°C (400°F) hotter than when the glaze first begins to melt.

The temperature limits for growing crystals are wide enough that most potters can expect some degree of success. However, if you wish to grow a specific shape or size, you must know your kiln well and determine your parameters through careful testing and record keeping.

PREVIOUS PAGE LEFT Close-up of crystal glaze © Matt Horne.

PREVIOUS PAGE RIGHT Close-up of cobalt crystal glaze © Glenn Woods.

BELOW LEFT Close-up of crystal glaze © Jose MARISCAL Paneque.

RIGHT Close-up of crystal glaze © Jose MARISCAL Paneque.

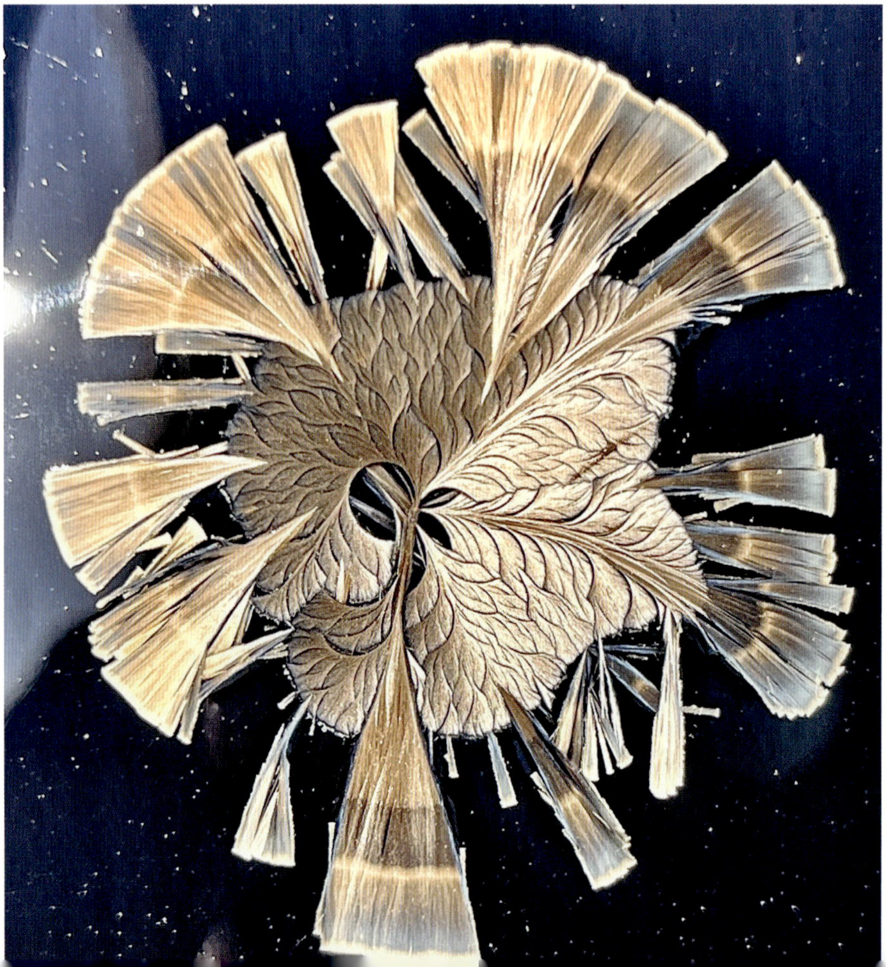

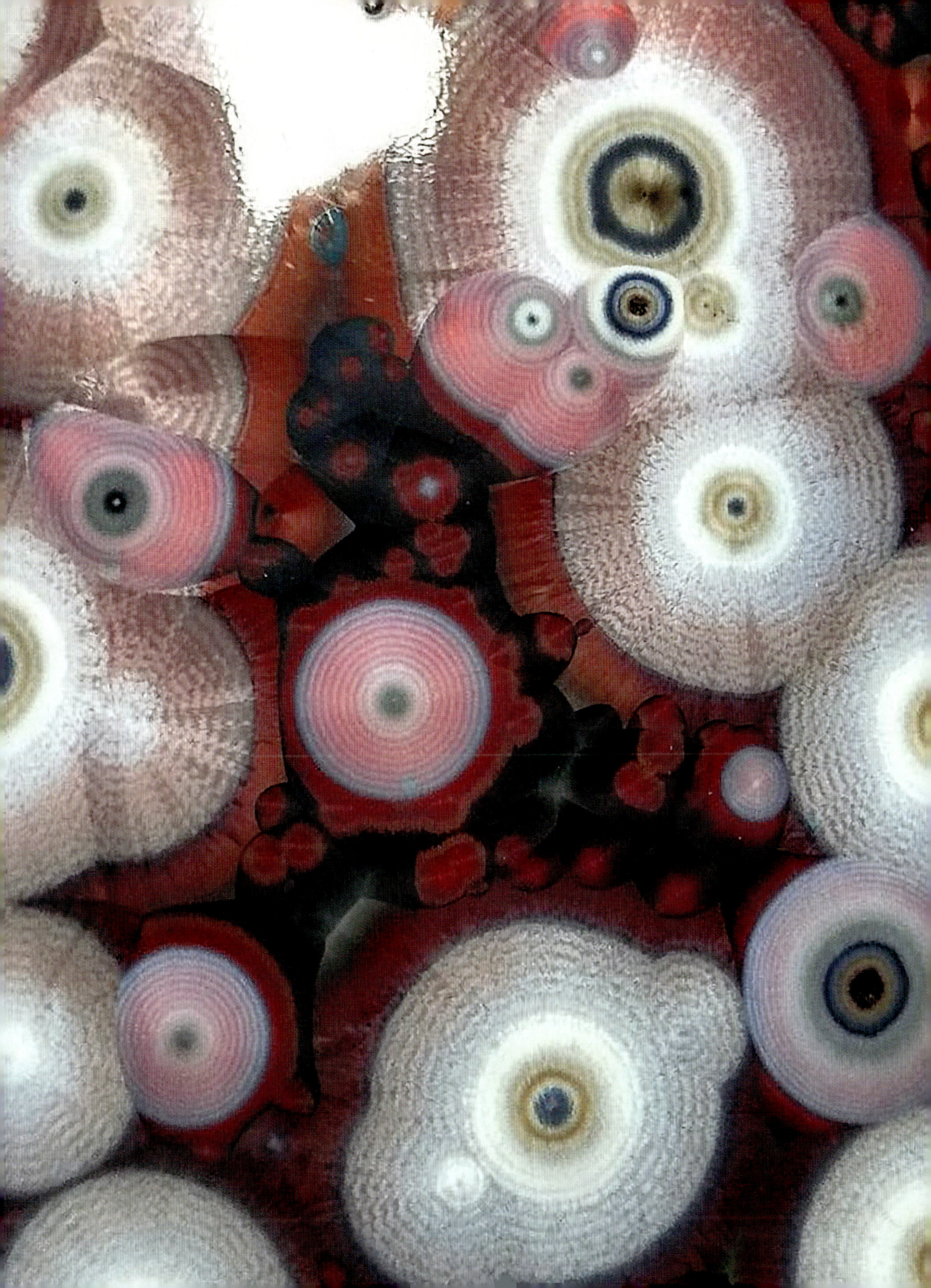

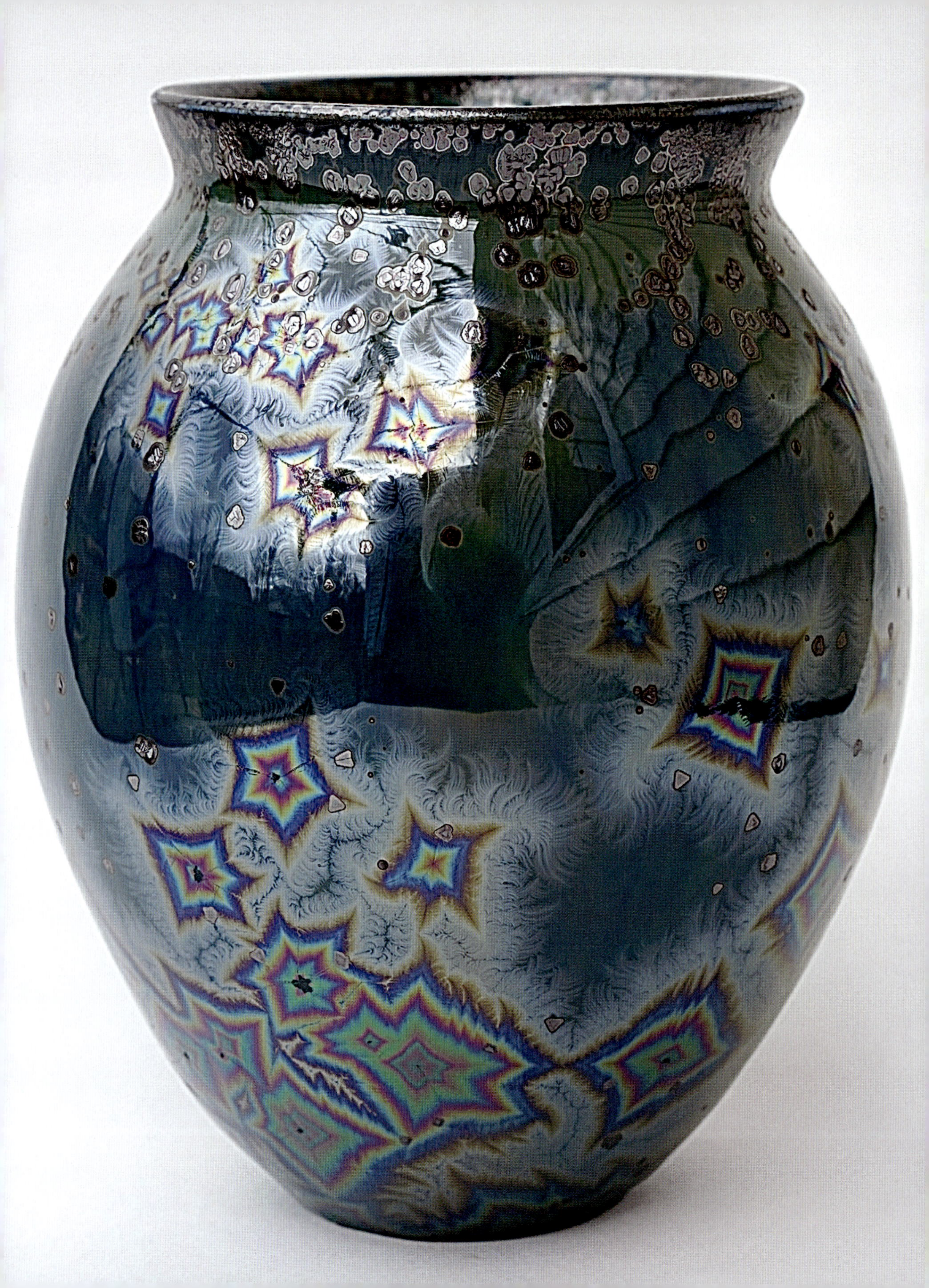

4 Types of crystalline glazes

LEFT Molybdenum glaze © Casey Anderson.

Crystalline glazes fall into two main groups: micro and macro. Micro glazes contain crystals that completely cover the pot and are usually so small they are difficult to see. Macro glazes have fewer crystals that can be easily seen. I will be focusing on macrocrystalline glazes.

These glazes produce a large variety of distinct crystal shapes of various sizes. They may cover the entire surface of the pot, or be limited to a few individual shapes. They can be divided into several categories.

Mid-range crystalline

Crystalline glazes fired to cone 6 produce smaller crystals than at the higher temperatures. To get these glazes to flux at these lower temperatures, frits usually comprise more than 50 per cent of the glaze recipe with secondary fluxes, such as smaller amounts of lithium carbonate or boron compounds, also assist with better melting.

Molybdenum

Molybdenum crystals are quite subtle and appear to float on the surface of the glaze. They are usually smaller than the zinc silicates and have angular edges, more like a star or a diamond. They often separate light into different-coloured rays, giving a rainbow effect. Sometimes the pot must be tipped to catch the light for the crystals to be fully appreciated.

These glazes are complicated, elusive and unpredictable, sensitive to temperature, and sometimes result in no crystals at all because the kiln was either slightly too hot or too cool. More successful results from molybdenum usually occur if several pots with the same glaze are placed next to each other in the kiln. Molybdenum vaporizes at 700°C (1,292°F) and settles on or fumes the nearby surfaces.

Metal oxides are often added to a molybdenum glaze to give colour to the matrix of the crystal. Molybdenum is used in amounts of around 5 per cent. Using smaller quantities in a zinc silicate glaze promotes the formation of other types of crystals, sometimes resulting in the glaze becoming opalescent. The glaze firing programme is similar to a regular crystalline firing.

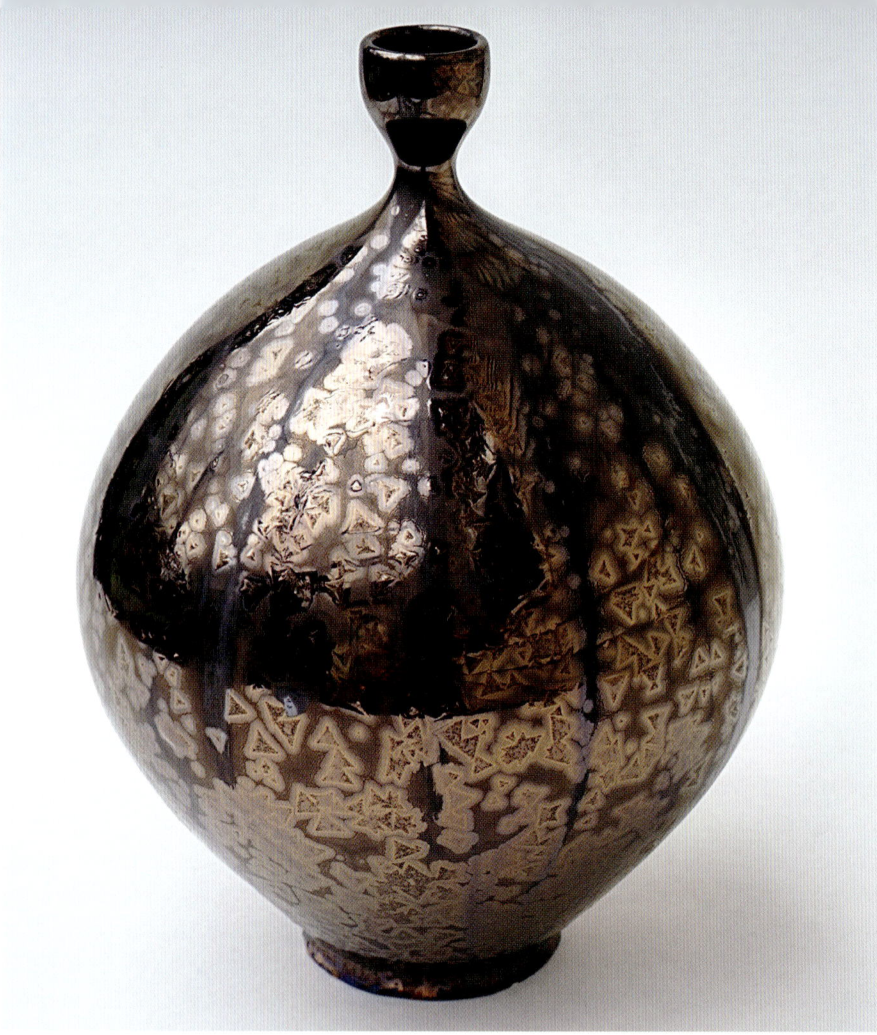

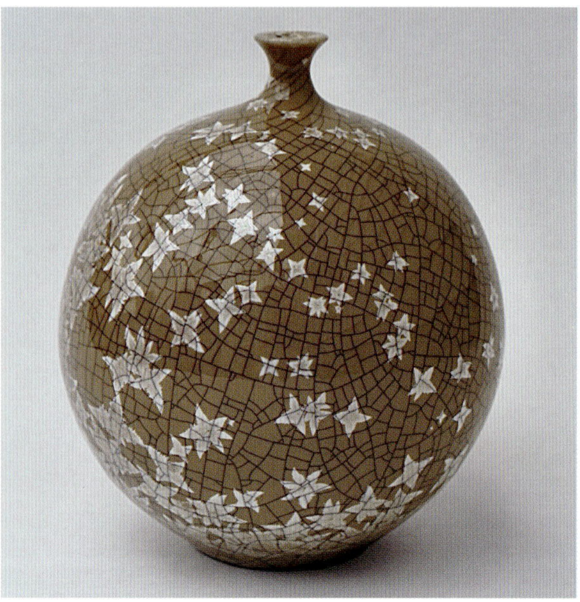

ABOVE AND LEFT Molybdenum glaze © Casey Anderson.

Aventurine glazes

Aventurines glazes differ from other crystalline glazes because they contain tiny, sparkling crystals that form when an excess of one or more metallic oxides have dissolved in the glass magma during the firing. As the kiln cools, the capacity of the glassy liquid to hold its dissolved material in solution decreases. The excess is deposited as small, suspended crystals which shimmer and sparkle as light hits them. A more complete description of aventurine glazes and their recipes is given in Chapter 9 (p.65).

Large individual macro

The crystalline glazes most of us are familiar with contain large, well-developed, distinct crystal patterns immersed in a glassy matrix. These are what most crystalline potters desire. While the crystals sometimes reach a size of 7.5–10 cm (3–4 in) across, they are usually much smaller, but achieving interesting background crystals or unusual patterns are also considered accomplishments.

These glazes are fired in the higher temperatures, cone 9–11. This allows most of the zinc silicate nuclei to dissolve and the crystals to grow from the remaining growth medium. After reaching maximum temperature, the kiln is cooled either by shutting it off or using a predetermined cooling rate, until it reaches the crystal-growing range. Since kilns differ, these temperatures are variable.

By holding the temperature at the point where the glaze is still slightly molten but just beginning to stiffen, the crystals develop, their size determined by the amount of time they remain in this state (typically three to five hours). Then the kiln is shut off and cooled naturally. The crystal shapes are determined by both time and temperature.

RIGHT © Ginny Conrow.

NEXT PAGE LEFT Bottle, base glaze with manganese dioxide added.

NEXT PAGE RIGHT © Mariela Tiferes.

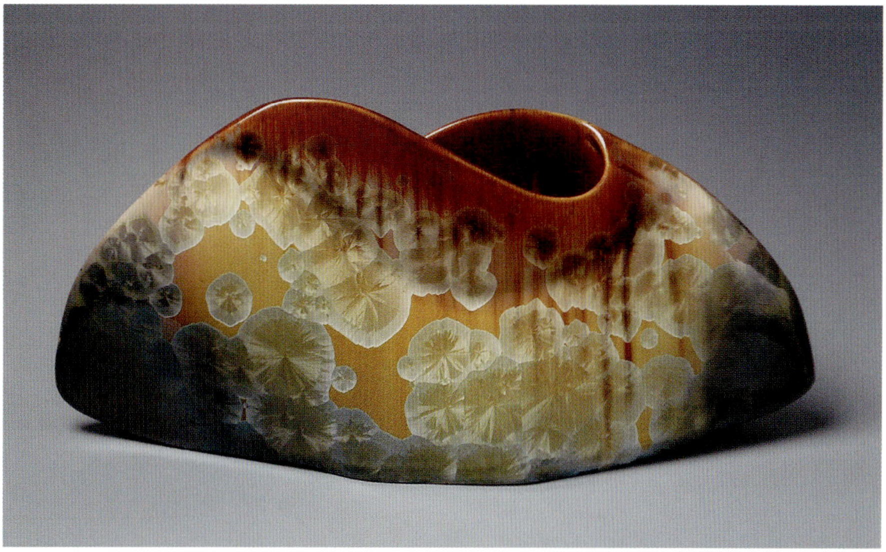

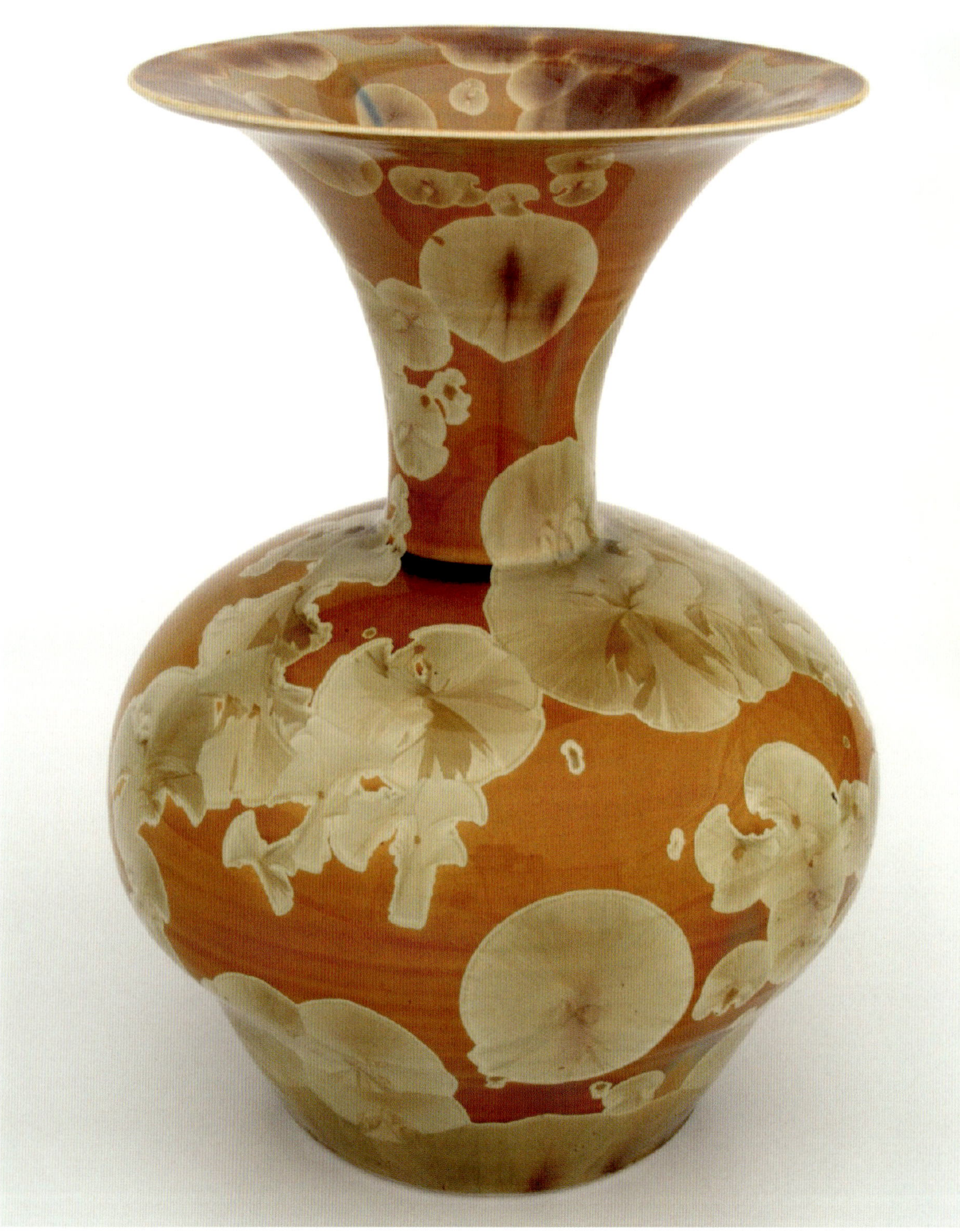

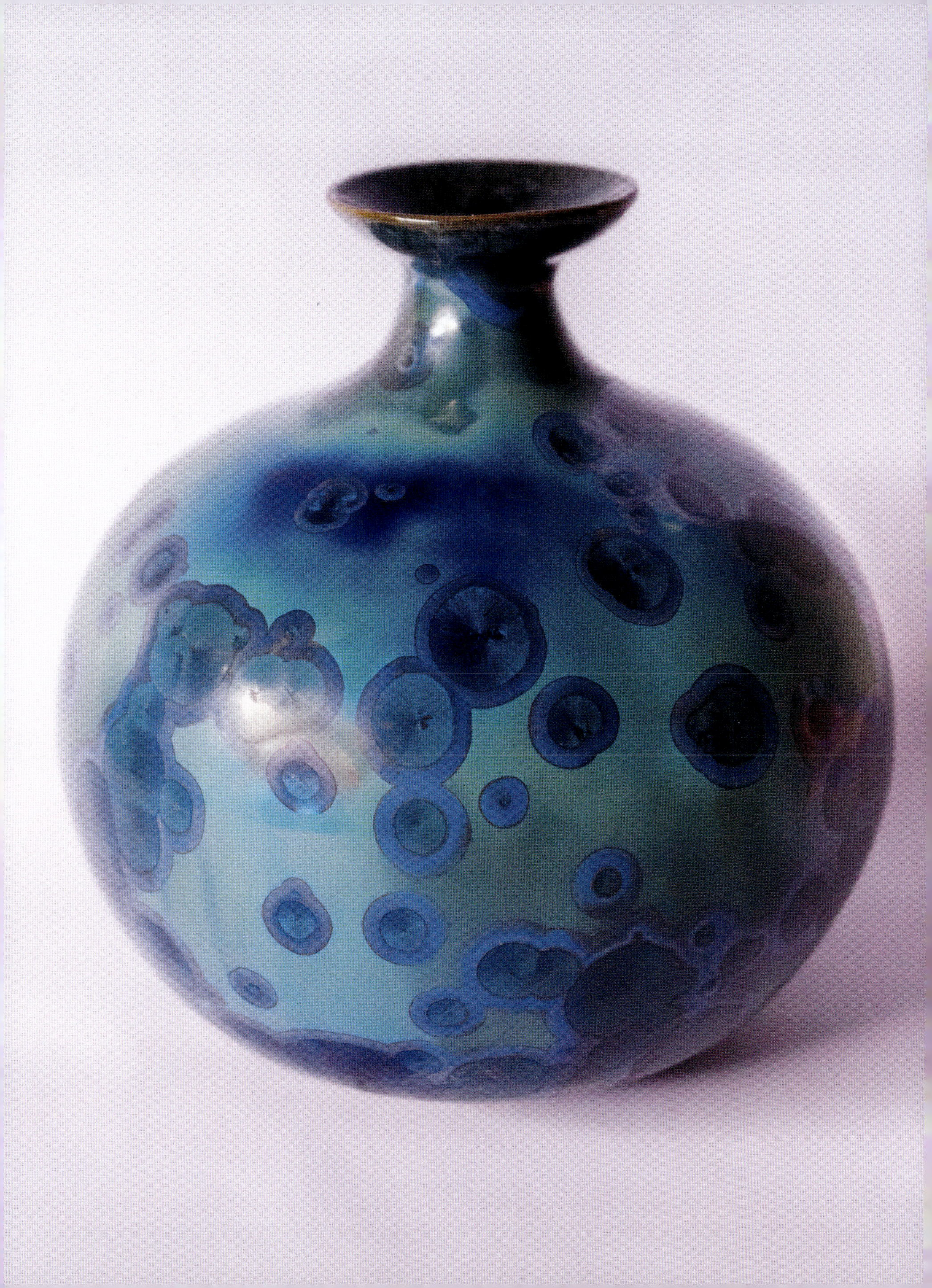

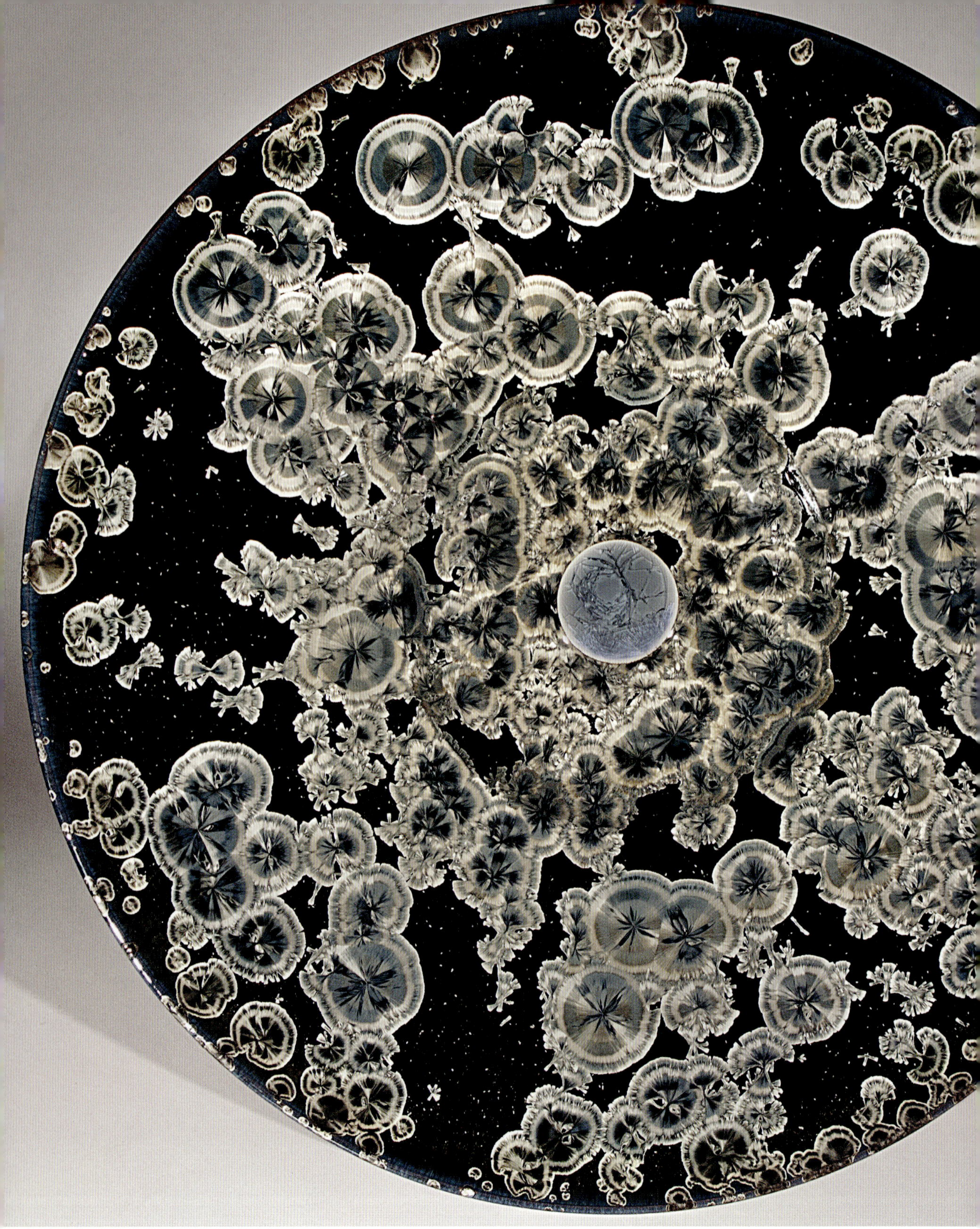

5

Suitable clay for crystalline glazes

Porcelain, characterised by its whiteness and smoothness, is the preferred clay for crystalline glazes. There is a quality to porcelain that is not apparent in other clays, which enhances the richness and vitality of crystallines, although that is not to say that other clays cannot be used.

Stoneware is suitable for matts and aventurines, but with macrocrystals, darker coloured clays show through the transparent glaze, influencing and darkening the background. Some materials added to stoneware, such as large-grained silica, ilmenite, molochite, or grog, can become nucleation sites resulting in too many crystals. If there are large-grained materials in the clay, when trimmed the scratch marks will be obvious, and the raised surface of granular material can provide nucleation sites. The clay should be smooth.

Potters Pat Geenty (UK) and Bill Boyd (Canada) use white stoneware for their larger forms. When the pots are leather hard, they spray two coats of porcelain slip onto the greenware. This allows them to work large and not be concerned about excess nucleation sites and the clay colour influencing the glaze.

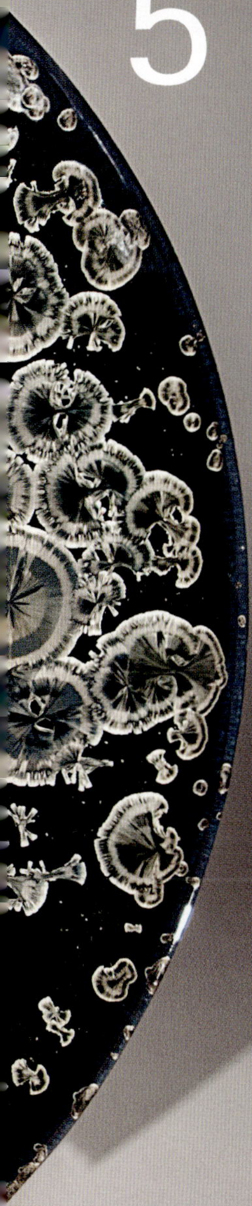

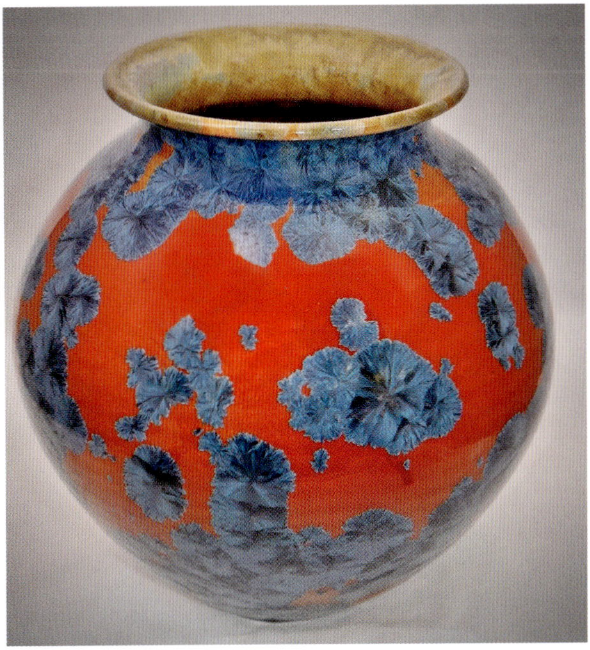

FAR LEFT © Bill Boyd.

LEFT © Pat Geenty.

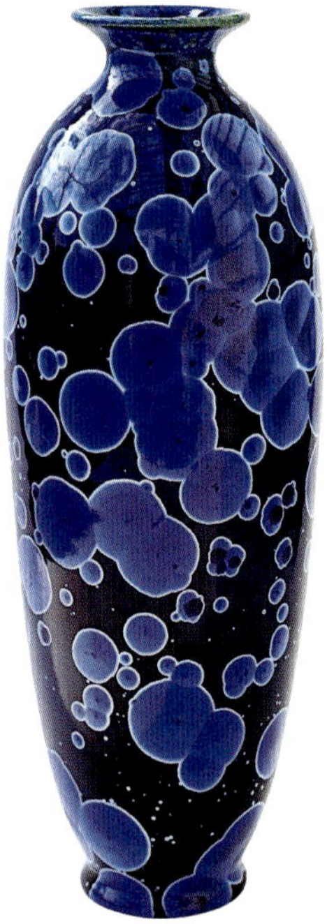

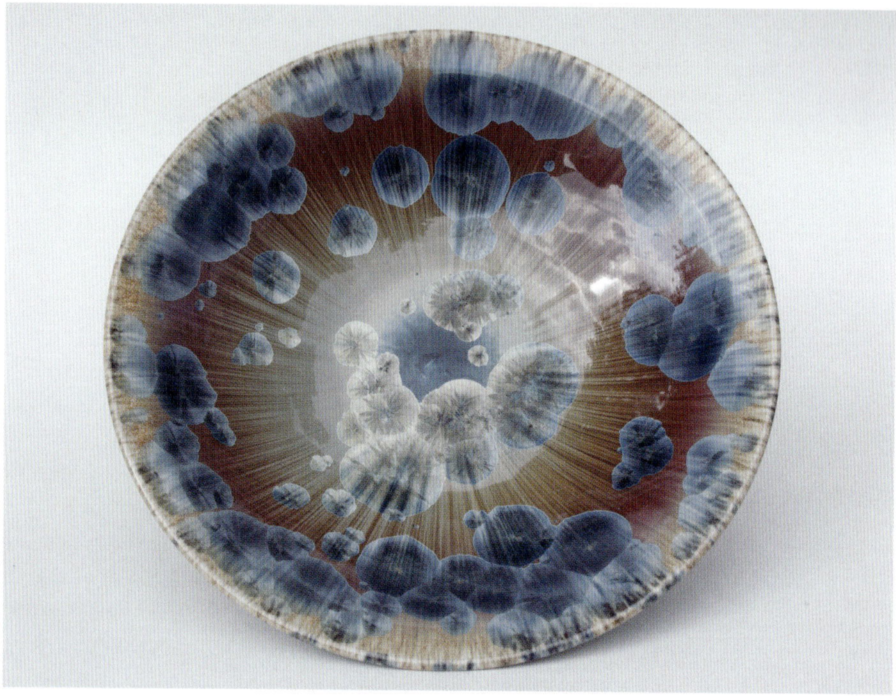

TOP LEFT © Casey Anderson.

ABOVE © Jamie Kozlowski.

RIGHT © Andrew Krueger.

Porcelain, however, doesn't contain contaminants that would influence glaze colour or crystal formation and provides a smooth, unrestricted surface. Since the pots are made thicker to absorb the thicker glaze application, the translucency of porcelain is not an issue.

These glazes are acidic. If the clay should start to flux during the melt, the alumina in the clay could be absorbed by the glaze, causing it to become matt. A porcelain that is slightly immature at peak temperature prevents the surface of the clay from being dissolved by the glaze.

Porcelain is usually made of 40–50 per cent non-plastic materials – namely silica and flux – making it less malleable and more difficult to work with than other clays. Therefore, various ingredients such as ball clay are added to increase plasticity, making it a little more workable and allowing a potter to throw larger pots with greater ease than with the translucent porcelains. Letting the clay age also increases its plasticity. Porcelain made in a slurry and then dried to a workable state becomes a better throwing clay.

Clays from different regions of the world are dissimilar and are forever changing as they become extinct or mines close. Before deciding on a clay body, run a series of tests on several different compositions, as some clays are easier to work with and produce nicer results with crystals than others.

Rather than give a list of clay recipes, my suggestion is to consult with a clay supplier and discuss your needs. Explain the forms and size you work at and the temperature you fire to. A reputable clay supplier will be able to recommend a clay body to suit your needs. Better yet, ask for samples of several clays and try them out.

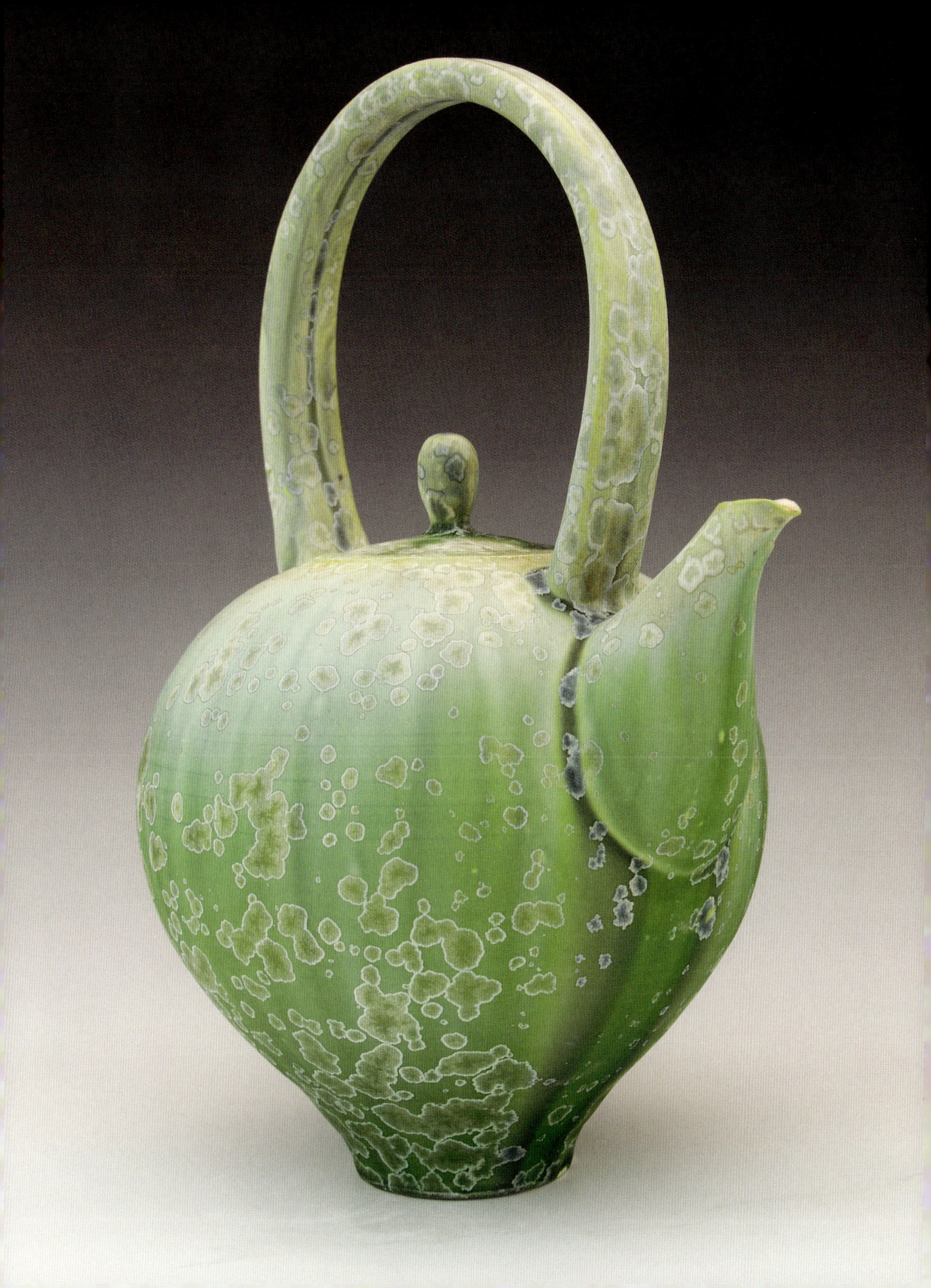

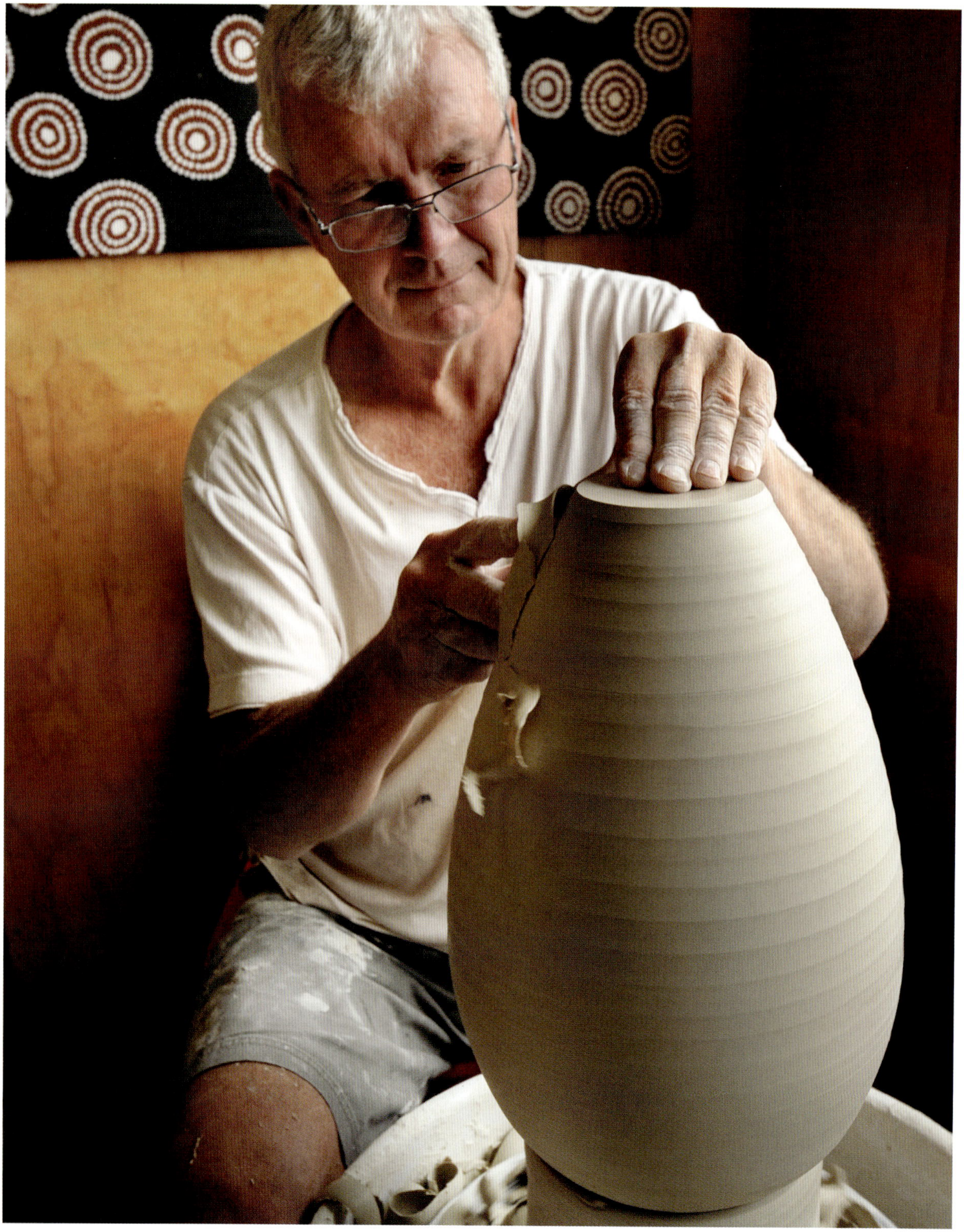

6

Forms, catch basins and test tiles

LEFT Ted Secombe trimming a tall pot © Ted Secombe.

Forms

Form is extremely important for every type of pottery. In fact, the structure of the vessel can be even more significant than the surface. The shape needs to highlight the glaze, not detract from it. A beautiful glaze on an unattractive or unsuitable pot does not justify the piece. Because of the richness of crystalline glazes, the form itself should be elegant and graceful. This is not the kind of glaze to put on a bean pot.

Throwing lines, appendages, or other adornments distract. Since the glaze is full of detail, surface decoration should be kept to a minimum. Adding further embellishment often makes the result too 'busy'.

Not all shapes are appropriate for crystallines because the glaze will collect and pool in pockets or crevices. A cylinder form is preferable and the tear-shaped bottle a favourite, its simplicity showing off the crystals to their best advantage. If the opening is small enough, the interior need not be glazed; alternatively, a conventional glaze may be used if the inside is visible.

Vases and jars work well because of their convex outline, upright height, and elegant profile. Their structure allows the glaze to run smoothly down the walls in an even application, without pooling. But, with vertical pieces, an allowance for the flowing must be considered. Since the foot of this shape is usually narrow, a catch basin underneath is used to collect the overflow glaze.

Bowls and plates are also suitable shapes, although crystalline glazes tend to pool on the lowest point, sometimes to a depth where it may look rough or resemble a puddle of glass. Therefore, glaze application should be thicker at the edges and thin towards the centre. By coating the underside with a standard glaze, no precautions for running need be taken.

With a form such as a covered jar, care must be taken to ensure the application is thinner near the edge of the lid, so drips don't form. A conventional glaze can be used on the interior.

Since aventurines are not as fluid as macrocrystalline glazes, usually no precautions against running are needed.

Catch basins

Although catch basins are not part of the finished piece, they are essential to protect kiln shelves from the running of these glazes.

In the book *Glazes for Special Effects*, Herbert Sanders suggests making a pedestal from porous, insulating kiln brick (#2400 brick for firing to cone 10) which is cut with a saw and filed or ground to the exact size as the pot's base.[18] The pedestal is then given 2-3 coats of kiln wash made from 50 per cent china clay and 50 per cent silica. Another ½-inch thick brick slab, coated with kiln wash goes under the pot and pedestal. This method, however, can be problematic: the glaze is acidic and eats into brick, sometimes causing the pot to topple.

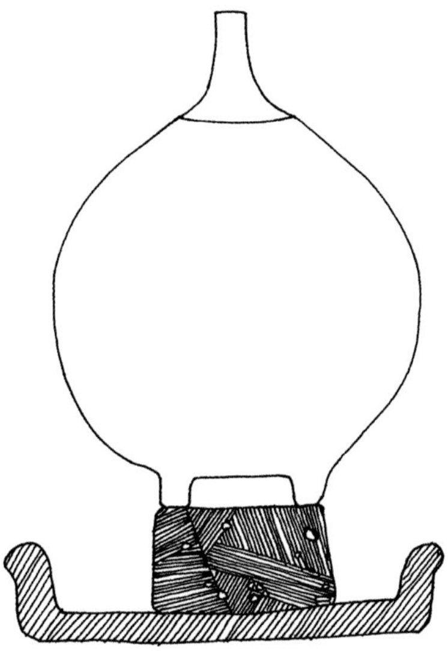

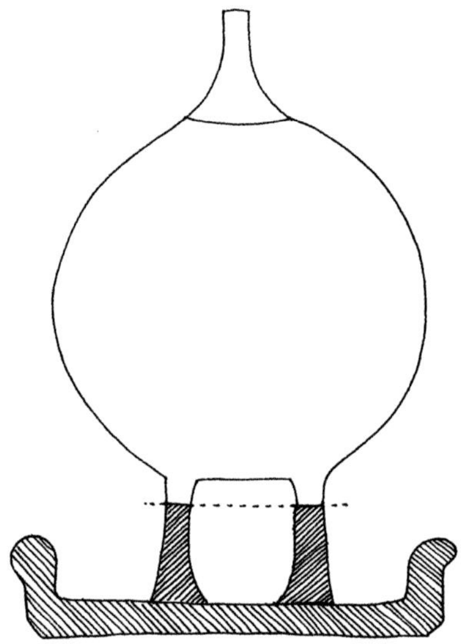

FAR LEFT Pot on firebrick.

LEFT Pot on pedestal in collection dish.

A simpler, more dependable method is the one that most potters now use. A pedestal about 1–2 inches in height that fits the bottom foot of each pot is made from the same clay as the pot, so there is no shrinkage differential. During the firing, excess glaze runs down the pedestal into a clay catch basin. Some potters make the pedestal and dish separately while others prefer to make them as one piece. Regardless, the sides of the dish should be high enough to contain excess glaze but not so high as to obscure the line where the pot bottom meets the top of the pedestal. The seam must be accessible in case a chisel or propane flame is needed to release the pot from the join. On smaller pieces, I make my pedestals separately from the dishes. That way, after I glue them to the pot bottom, I have a handle to hold when I dip the pot into the glaze. For larger items, I make the pedestal and basin in one piece and either spray or brush on the glaze. A thicker dish survives better than a thinner one; however, the inside bottom must be flat so the narrow pedestal supporting the pot won't topple.

[18] Herbert H. Sanders, *Glazes for Special Effects*, Watson-Guptill Publications, New York, 1974, p.39.

RIGHT Fired pot on catch basin © Jamie Kozlowski.

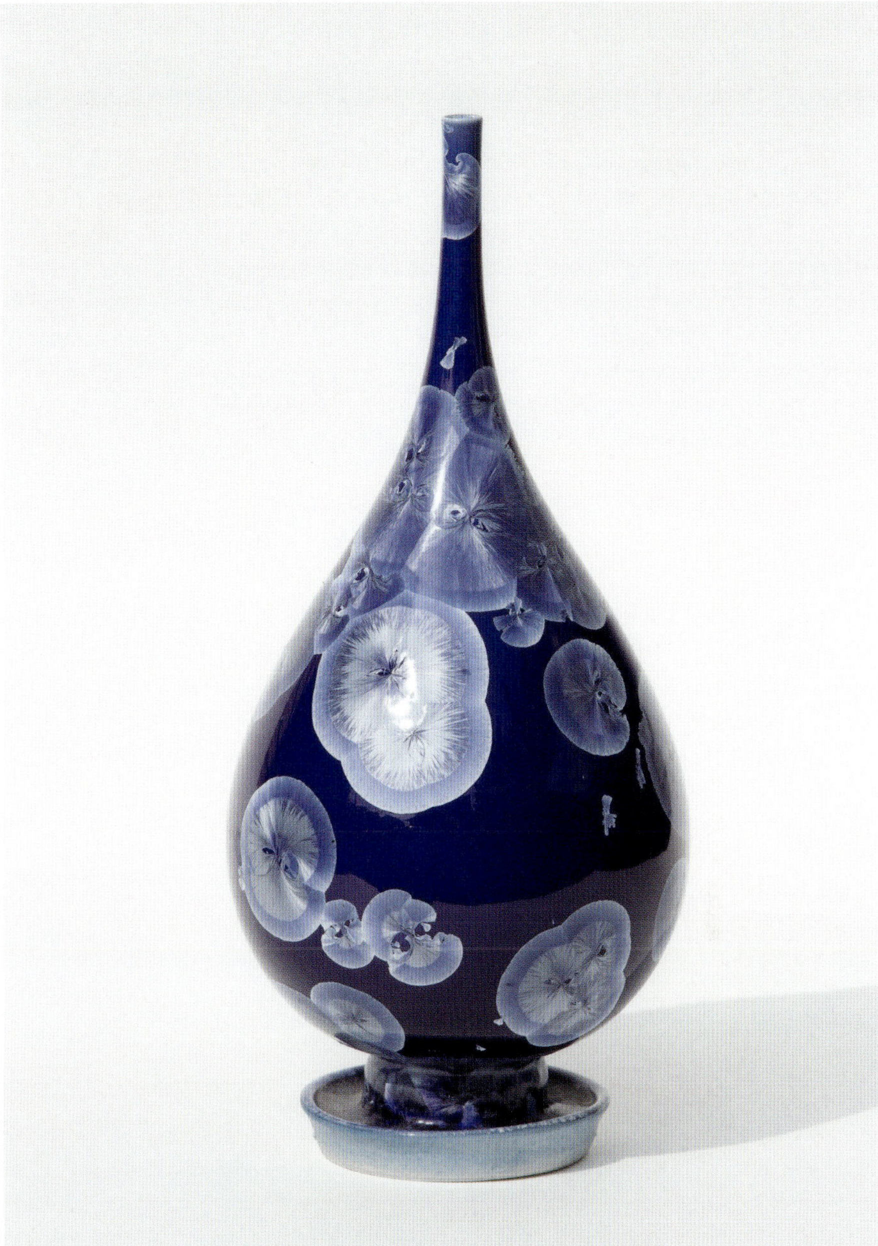

The pot's foot rim and the top of the pedestal must have flat, sharp edges to allow close contact between the two, otherwise, glaze may run under a rounded edge and make removal more difficult.

I trim all foot rims to a template and then make the top of the pedestals the same size. I usually trim the bottoms of the pots to three distinct sizes: smaller pots to one size, larger to another, and larger still to another. I make the same number of pedestals and dishes as there are pots in that particular size, and then I make a few extra pedestals. After the bisque firing, I store them in separate boxes according to size. Using a template ensures the bottom of the pot and top of the pedestal fit, and having a selection of same-sized pedestals for similar-sized pots certainly speeds things up. It is more labour intensive to make a pot, trim it, and then make a separate pedestal to fit.

Larger pots are usually one-of-a-kind, so I make a one-piece catch basin. To ensure an exact fit, I place the basin under the trimmed pot and trim it to fit.

After bisque firing, the bottoms of the pots and tops of the catch basins are flattened with emery cloth or placed on a grinder for a few seconds. The two pieces are then joined with a paste of 50/50 china clay and alumina (or kaolin) mixed with a little white glue and a small amount of water. The mixture is applied thickly to the top of the pedestal and the two pieces are fitted together. This usually results in the pot coming off the pedestal with a slight tap after the firing. In Chapter 15 (p.119), more detailed methods of removing pot from pedestal are described.

BELOW LEFT © Yadollah Ziarati.

RIGHT © Matt Horne.

Test tiles

To understand your glazes, you must conduct tests. Because these glazes flow, tests should be fired on vertical forms. Using the same clay as the pots, throw a cylinder about 10 cm (4 in) high and 25.5 cm (10 in) wide with a thick bottom. After it dries to leather hard, cut out the centre of the bottom, leaving about 5 cm (2 in) closest to the walls. Then cut the pot vertically into about ten sections. Each section becomes a test tile. The glaze is applied to the inside wall of the tile. If you drill a small hole in the top of each tile, you can later pass a string through all the tiles of one test batch, so they can be hung together.

See Chapter 10 (p.73) for how to glaze without using a catch basin.

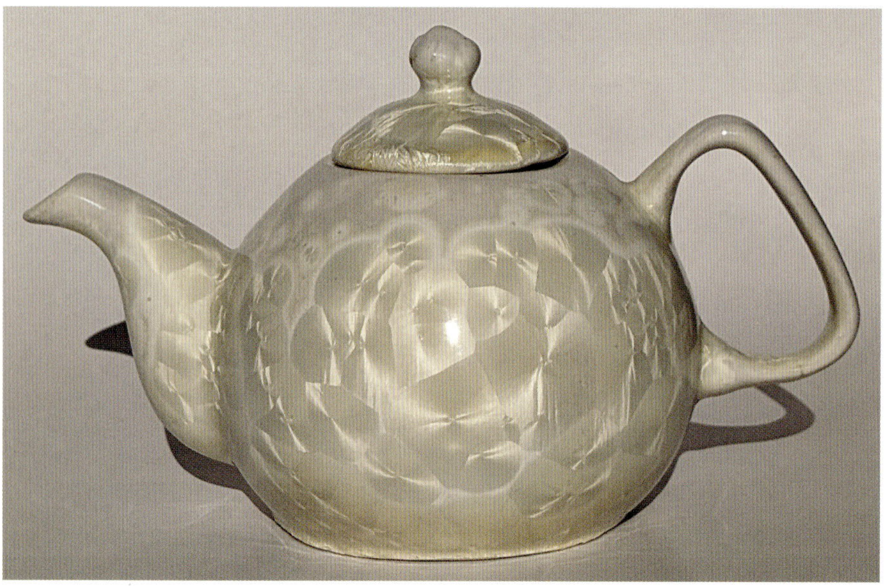

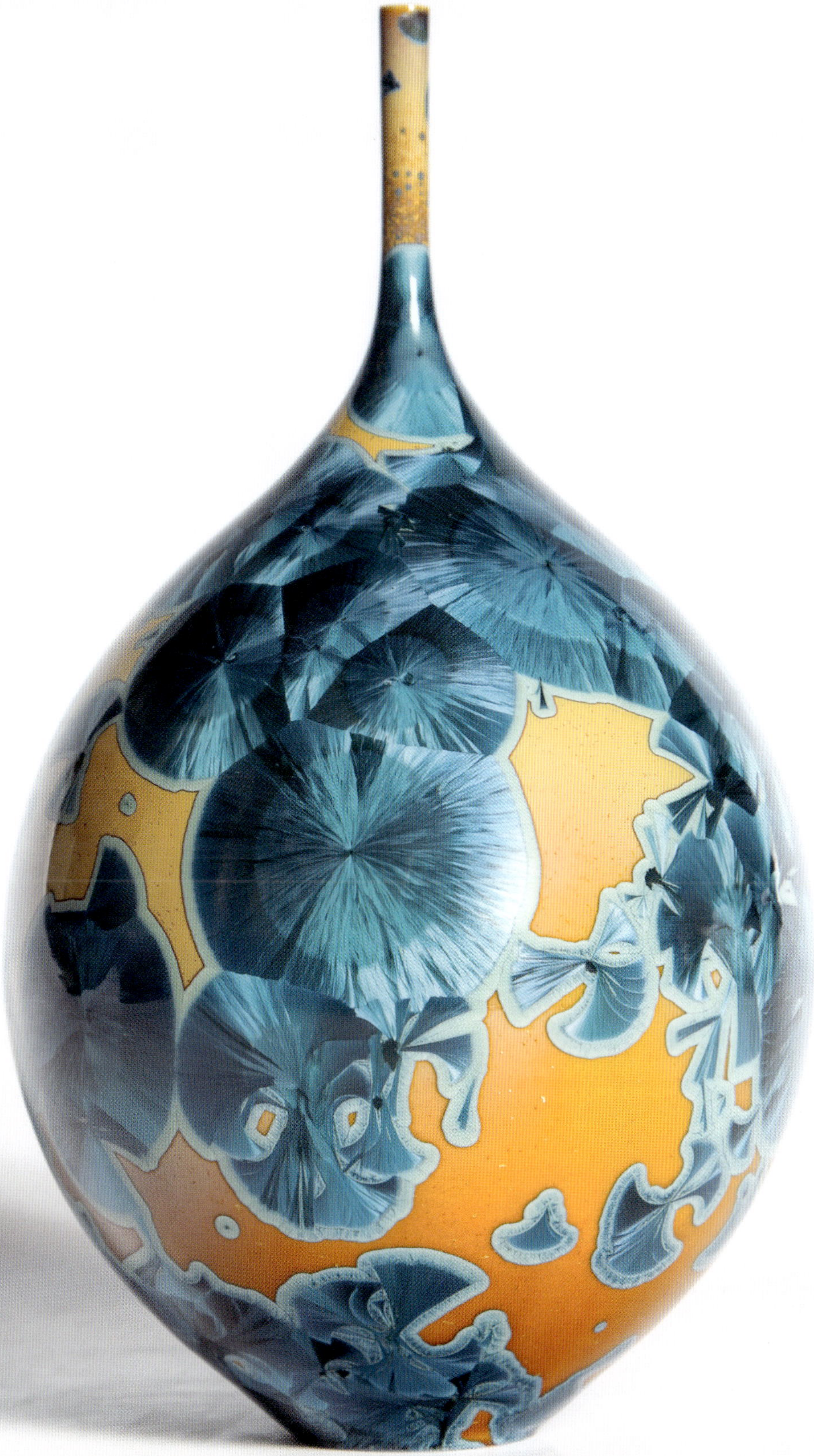

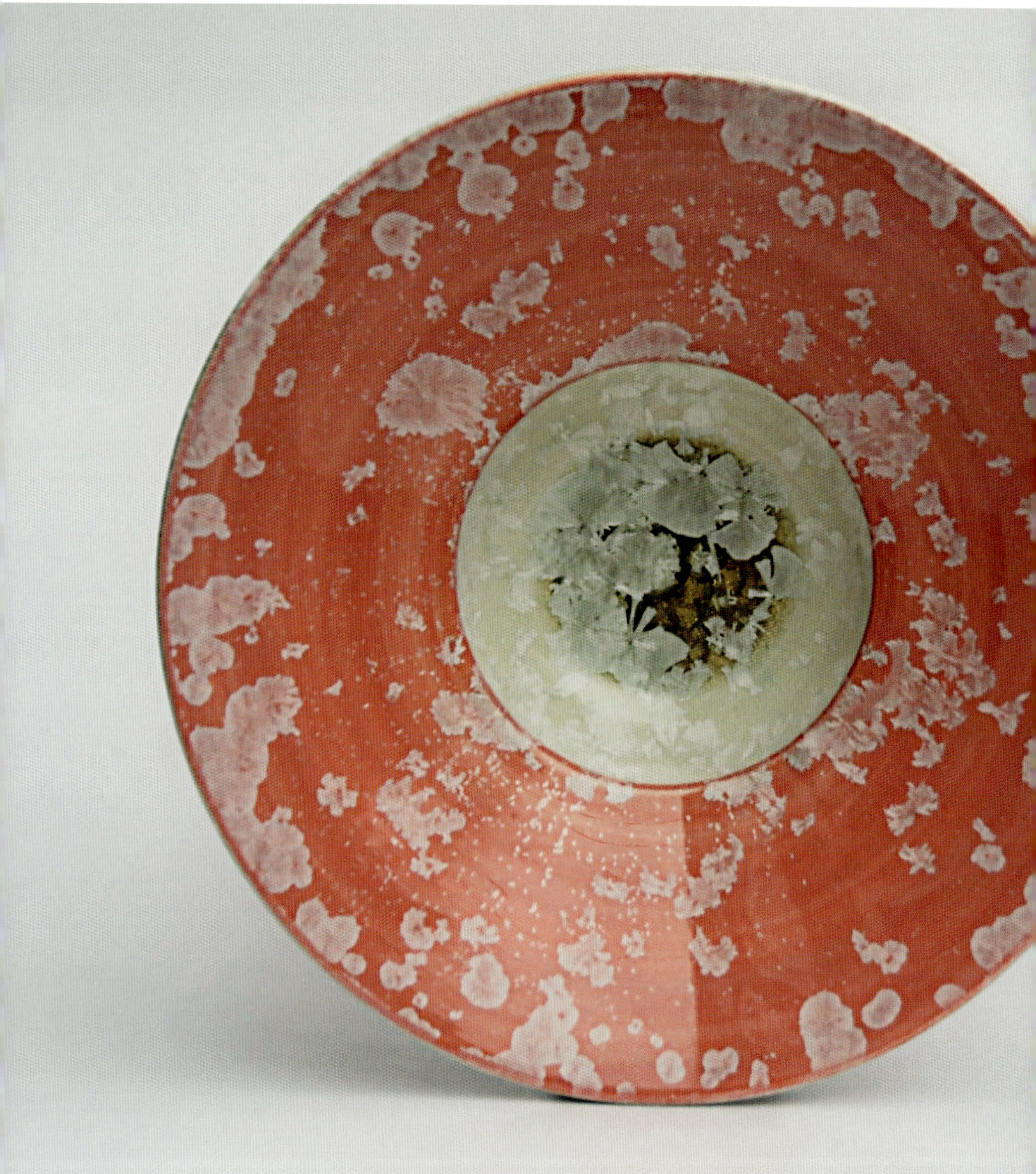

7 Ingredients in a crystalline glaze

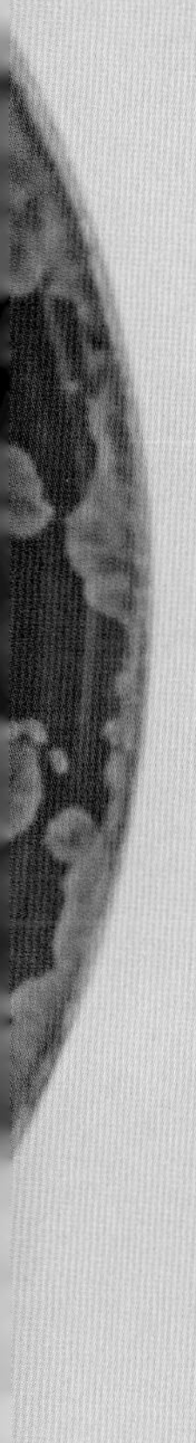

Both micro- and macrocrystalline glazes are usually made up of approximately 50 parts frit (assuming one is using a frit rather than raw ingredients), 25 parts zinc oxide, and 20 parts silica. Other materials, in lesser proportions, are added to alter the character of the finished glaze. Knowing glaze ingredients and how they react in the melt is essential for understanding crystalline glazes. All glaze recipes in this book are expressed as parts adding up to approximately 100 or 1,000. Recipes are continually adjusted by increasing or decreasing certain amounts, and for this reason, some of the recipes I share here don't total exactly 100. If you find it easier, you can convert these amounts to parts in 100 or 1,000, but in my experience, it makes little difference and isn't really necessary.

Silica

Silica is the main constituent of more than 95 per cent of known rocks. It is also called flint, or quartz in its pure crystalline state. When melted and cooled, it readily forms a glass. Silica also enriches the colour of glazes. Because it melts at an extremely high temperature, it must be used in combination with fluxes. Silica in a glaze should be at least 325 mesh.

Zinc oxide

Zinc oxide, at the middle and high temperatures, is an active flux. When used in excess in a glaze low in alumina, and then cooled slowly, zinc will produce crystals. Zinc and silica have an affinity for each other and readily combine to form zinc silicate (Zn_2SiO_4). Increasing the zinc content increases the number of crystals. However, too much zinc produces too many crystals, resulting in a matt glaze.

Calcining

Calcining is where a glaze ingredient is heated to the temperature at which chemically combined moisture is driven off. Burning off the organic soluble salts reduces shrinkage.

Calcining is relatively simple: Put the dry material in a bisque-fired bowl and take it to bisque temperature. The temperature is not critical. After heating, pass it through an 80s screen.

Almost all zinc oxide sold today is calcined, but check with your supplier to be sure. Zinc oxide can absorb moisture from the air, so this ingredient must be stored in a sealed plastic bag or other airtight container. When left exposed over time, atmospheric moisture causes zinc oxide to change and become zinc hydroxide (Zn(OH)2).

Weights given in glaze recipes are for calcined zinc oxide, which weighs less than uncalcined. If using uncalcined zinc oxide, or if the zinc has been left exposed to the atmosphere for a length of time, you may be adding too little of the ingredient, and the glaze will tend to peel off the pot as it dries. If it does peel, or crawling occurs, I recommend using a new batch of calcined zinc.

PREVIOUS PAGE Red stain on outer rim of plate, clear glaze over all.

Frit

A frit is a manufactured compound that has been fired to a molten state, cooled, then ground to a powder. Using a frit simplifies glaze preparation – rather than weighing out and adding several materials, a single, convenient material can be added. A frit lowers the overall melting temperature of the glaze ingredients and renders water-soluble and/or toxic ingredients insoluble by causing them to combine with silica and other added oxides. It also ensures that a homogeneous glaze coating remains on the surface of the clay, not in the water, and that no soluble material is absorbed by the clay where it could act as a flux. A crystalline glaze made without a frit must be used immediately because of the solubility of the materials.

When using lead compounds, commercial frits are formulated to balance lead in the correct ratio to silica and each frit is designed to balance at a specific temperature. This process supposedly renders the lead non-leachable and safe for commercial applications. However, correct firing temperatures must be reached to ensure no leaching occurs, which is often difficult to predict, and other material additives may alter the chemistry and prevent the locking in of all the leachable elements. When using a frit that contains lead, do not use it on the inside of a vessel that may be used to contain food, and be aware of toxic fumes given off during the firing. I will not be recommending any frits that contain lead.

Certain frits may be difficult to obtain or may be unavailable commercially. Before frits were commercially available, potters either used unfritted (raw) materials or made their own. To make a frit, first make a crucible (smelting pot) from high-alumina fireclay fired to cone 10. Weigh out and dry mix the ingredients to be fritted, put them in the crucible and place the crucible in the kiln. Heat the kiln until the materials melt, then pour the red-hot liquid into a tank of water. On contact with the water, the material quickly freezes and shatters into glass fragments, which are strained, then ground in a ball mill until the particles are small enough to pass through an 80s mesh screen.

This potentially dangerous and complicated procedure demands extra safety precautions, and an extremely long grind in a ball mill is required to pulverize the frit to a suitable fineness. Unless your studio has the facilities, and you have experience with the process, the safety measures and extra work required render this procedure impractical, and I do not recommend it. Most potters rely on the many commercial frits available on the market today.

Frits differ depending on the manufacturer. The most popular commercial frit used by crystalline glaze potters was Ferro Frit 3110, which at the time of writing is manufactured by Vibrantz Technologies. I will call it 'frit 3110' in glaze recipes. Fusion Ceramics Inc., an Ohio company, makes a frit called F-75 which is considered a substitute for 3110.

Names and availability of different frits vary geographically, so potters should try to work with the frits available in their region. It is a challenge to keep track of what frit a manufacturer makes and what name it goes by, so potters should know the formula of a particular frit and ask for a chemical analysis from either the manufacturer or their ceramic supplier. Sometimes, when a certain frit is unavailable, other slightly different frits may be substituted. Having a knowledgeable ceramic supplier can certainly make ordering frits easier.

Alumina

Alumina is added to increase viscosity, making the glaze less likely to run off the pot when melted. Used in small amounts, it helps the finished glaze adhere to the pot in a uniform thickness. Alumina levels must be kept low because it produces too many crystals, resulting in a matt texture. Alumina is usually introduced in the form of kaolin.

Feldspar

Feldspar is a source of alumina and a natural glaze at high temperatures. It is used to lower the temperature of silica and is considered a primary flux.

Whiting

Whiting, also known as calcium carbonate, acts as a flux particularly at mid to high firing temperatures. It adds to the durability, hardness, and surface quality of the glaze. Higher quantities can promote the formation of calcium silicate crystals, resulting in a matt crystal glaze.

Titanium dioxide and rutile

Titanium dioxide, used in amounts up to 5 per cent, encourages crystal growth. It also acts as a colour modifier, producing brilliance and purity. If the percentage of titanium dioxide is increased beyond 5 per cent, the glaze becomes matt. Rutile is an impure oxide of titanium and contains small amounts of iron. It is sometimes substituted for titanium dioxide, and gives a yellow cast to the glaze colour.

Other oxides

- Sodium, lithium, and potassium oxides are active fluxes, which dissolve other oxides and contribute fluidity to the glaze.
- Calcium, barium, and magnesium oxides are sometimes added to give lustre and elasticity.
- Boron reacts like alumina, impeding the growth of large, isolated crystals.

These oxides can be added directly to a glaze. More often, they are added to the base glaze in the form of a frit.

Colourants

Metals which act as colouring agents may be added to glaze recipes in small amounts, either individually or in combinations of two or more. Cobalt carbonate, copper carbonate, and manganese dioxide are the three used most often. Oxides of these elements produce stronger colours, while carbonates are more subtle. Rutile and the oxides of iron, chrome, nickel, and uranium also provide interesting results. Some colourants seem to concentrate more in the crystal than the background. Generally, as the amount of colourant is increased, the crystals darken and the background colour becomes more intense.

Colouring oxides in amounts up to 8 per cent, sometimes more, of the total dry weight of the glaze recipe can be added. When two or more colourants are used, concentrations of each should be reduced. Colourants can be included in the dry batch or mixed with a little water, passed through an 80s mesh screen and then added to the glaze.

The following is a list of colourants with their concentration ranges given as a percentage of dry weight. The colour each produces is as follows.

INGREDIENT	PROPORTION	RESULT
Cobalt carbonate	0.5–2%	Blue crystals on pale blue ground to dark blue crystals on dark blue ground
Copper carbonate	1–4%	Silver crystal on pale green ground to green on green ground
Manganese dioxide	1–5%	Pale lavender to pinkish tan to brown or purple
Manganese carbonate	2–4%	Pink (has a lower melt temperature than manganese dioxide)
Red iron oxide	1–4%	Yellow to green/gold crystals on brown ground
Nickel oxide	1–2%	Green with yellow background
Rutile (dark)	1–4%	Amber yellow

Experiment with percentages and combinations; all kinds of colours can occur. The possibilities are endless, but keep in mind that the results are also dependent on the other ingredients in the glaze. Additions of colouring oxides added to two different base glazes can give two very different results, but generally, here are a few examples of what can be expected:

INGREDIENT	PROPORTION	RESULT
Manganese dioxide Cobalt carbonate	1–2% 0.2–0.8%	Silver crystals on a tan background
Copper carbonate	6–8%	Medium green crystals with black rings on green background
Cobalt carbonate Copper carbonate	0.1% 0.3%	Clear blue crystals on cream background

Colour schemes are relatively uncomplicated until you have several oxides in the glaze. Once you start adding colourants, the melting temperature alters, sometimes by several degrees.

Stains

Some colours such as pink, peach, lavender, and purple are difficult to achieve in a crystalline glaze. Commercial stains may be an answer, but they require extensive testing because ingredients in a stain may enhance or discourage crystal growth.

Andrew Krueger (USA) makes a slip using his powdered clay body mixed with a 15 per cent stain inclusion slaked with minimal water and deflocculated to a sprayable consistency. This is sprayed on bone-dry greenware. Spraying gives a consistent, relatively thin application. Brushed-on slip is difficult to apply evenly and often leaves visible brushstrokes. Dipped or poured application is challenging with fragile greenware.

BELOW LEFT AND BELOW Glaze over red stain © Andrew Krueger.

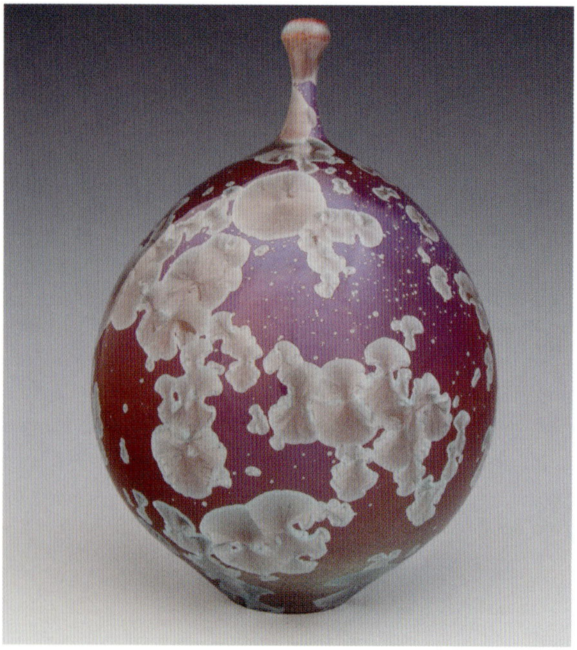

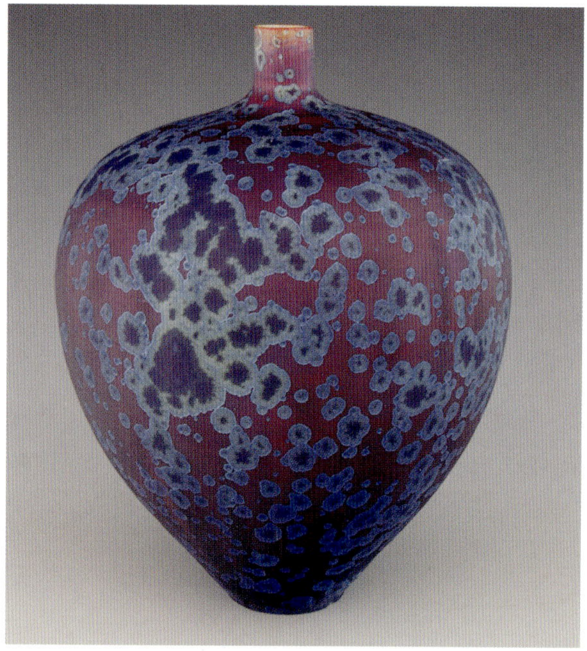

Jamie Kozlowski mixes his stain in a blender, and if well mixed, sprays on greenware without straining. The CMC (carboxymethyl cellulose) gum and the soap seem to help with that.

Jamie's recipe

INGREDIENT	MEASURE
Dry porcelain	1000 g
Water	1500 ml
Stain	10 %
CMC gum	2 g
Dawn® dish detergent*	1 drop

*Different detergents react differently, so if you choose to use another dish soap, experiment first.

RIGHT Glaze over red stain © Jamie Kozlowski.

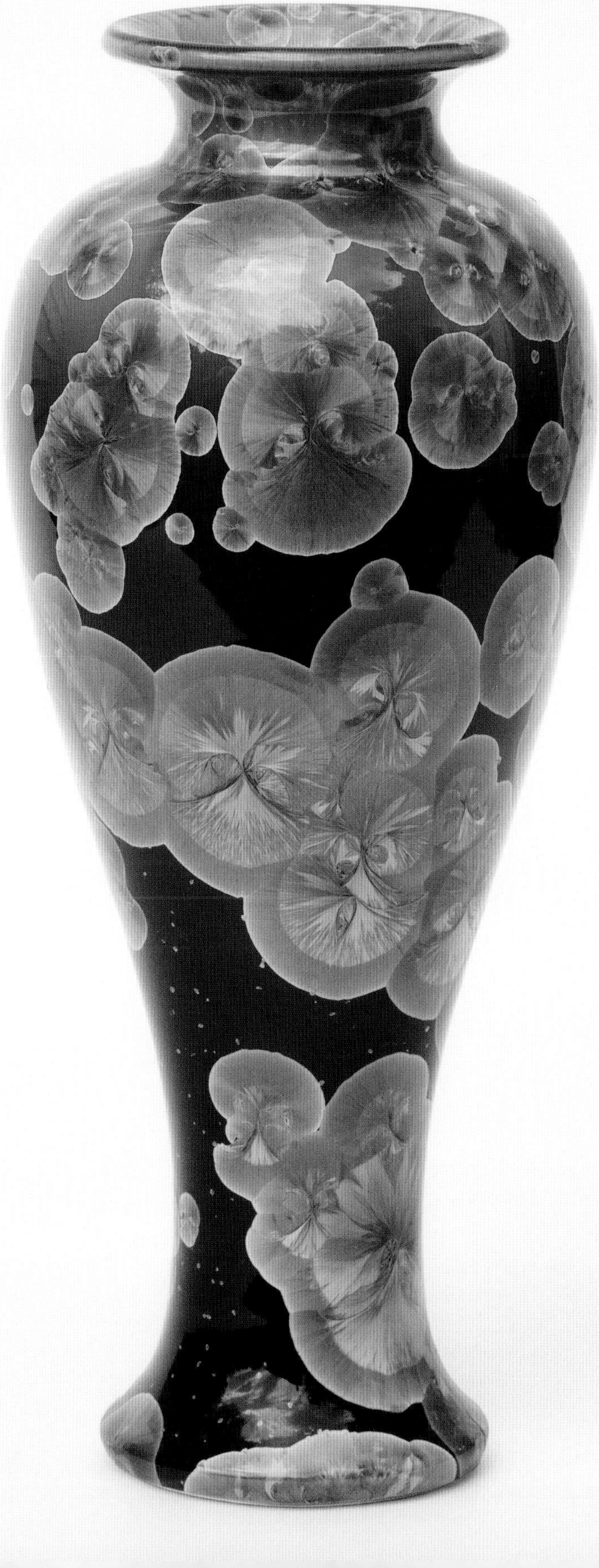

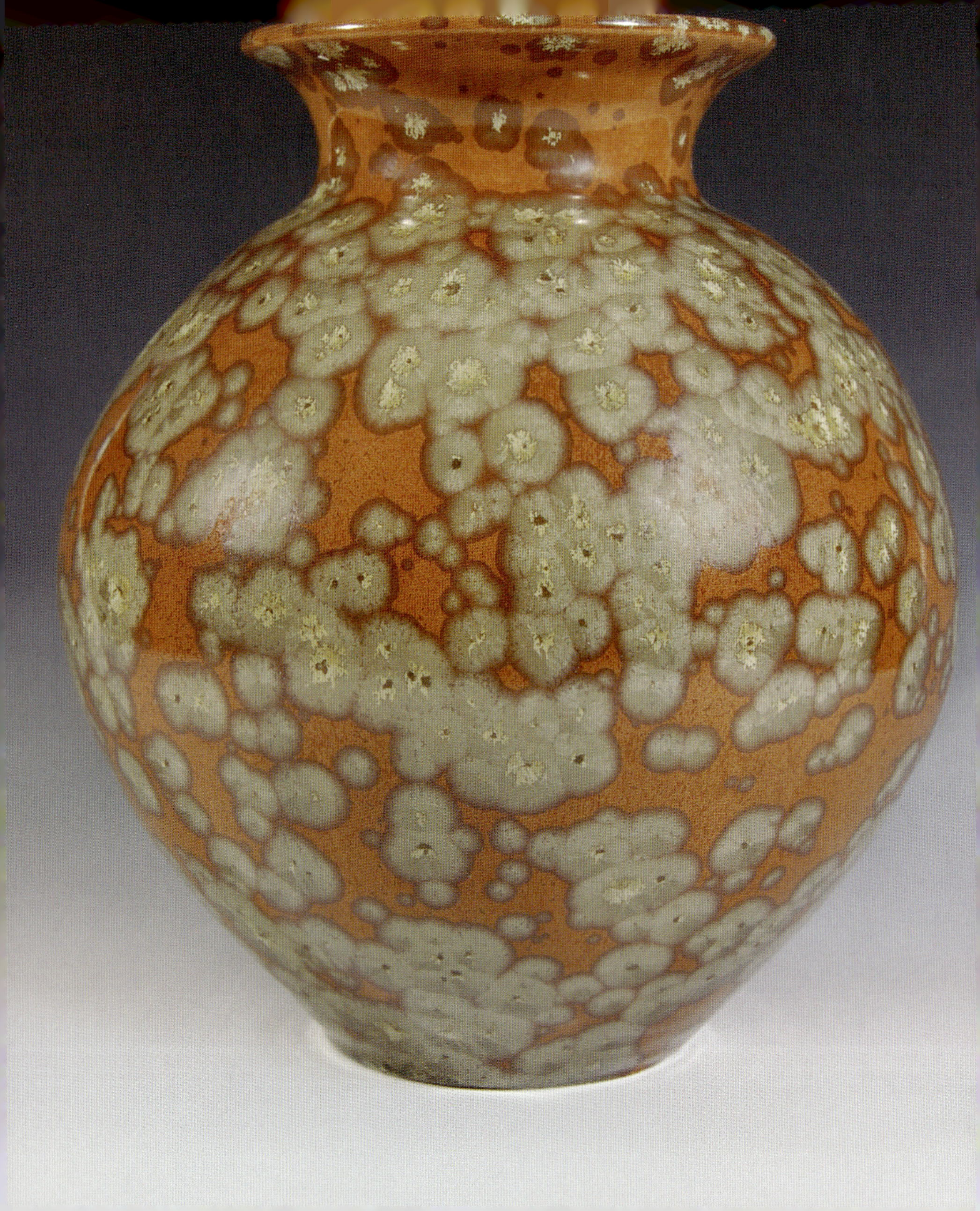

8 Glazes

LEFT Cone 6 glaze
© William Schran.

Crystalline glaze recipes are posted daily on the internet, so it seems redundant to give a lengthy list of them. I will, however, give a few base glazes in the different temperature ranges. This could be a starting point to developing your own. How to mix and apply these glazes is explained in Chapter 10 (p.73).

If you are just starting to work with crystalline glazes, begin with one base glaze, get to know its characteristics, how colourants affect it, and how it behaves in your kiln before trying another. When you're testing, only alter one amount or material at a time. If you alter the firing schedule, do not modify the glaze until you see what difference the firing has made.

The same glaze will give different results under different conditions: clay used; variables in glaze ingredients; colourants; temperatures; speed of firing; kiln atmosphere; and whether fired in oxidation or reduction. Listed here are some base glazes. The cone number beside each represents the maximum kiln temperature.

Mid-range crystalline glazes

Optimal crystal-growing temperatures for cone 6 glazes appear to be in the 1065–1049°C (1950–1920°F) range.

Cone 6

INGREDIENT	PARTS
Frit 3110	50.00
Silica	23.00
Zinc oxide	23.00
Lithium carbonate	4.00
Total	100.00

Cone 6

INGREDIENT	PARTS
Frit 3110	50.00
Silica	14.00
Zinc oxide	27.00
Talc	5.00
Spodumene	4.00
Total	100.00
Add: titanium dioxide	2%

Cone 6 (Bill Schran)

INGREDIENT	PARTS
Frit 3110	51.00
Zinc	25.00
Silica	19.00
Dolomite	5.00
Lithium carbonate	3.00
Black nickel oxide	1.00
Red iron oxide	4.00
Manganese dioxide	4.00
Total	112.00

Macrocrystalline glazes

The following glazes all contain frits. The frits given are readily available and are non-toxic. The following was first posted in *Ceramics Monthly* magazine in 1975 by David Snair.[19] Most macrocrystalline glazes today are variations of this recipe, it is the base used by many potters today and can be a starting point when making your own glazes.

Cone 9/10 (David Snair)

INGREDIENT	PARTS
Frit 3110	48.40
Zinc oxide	24.35
Silica	17.95
EPK	1.52
Titanium dioxide	7.78
Total	100.00

[19] David Snair, 'Making and Firing Crystalline Glazes', in *Ceramics Monthly* 23/10, 1975, pp.21-26.

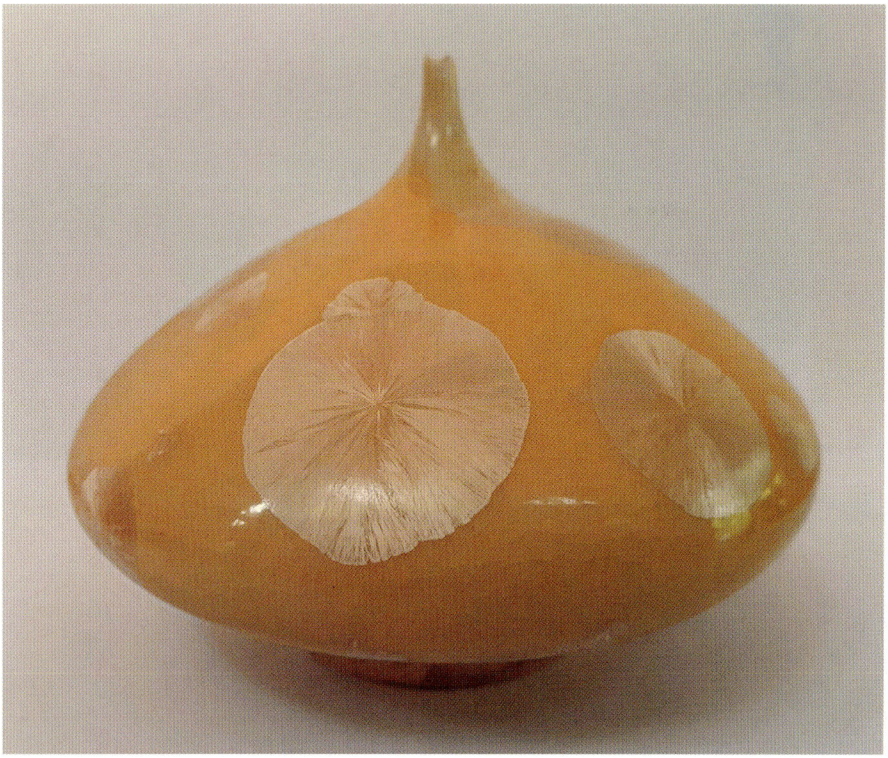

RIGHT David Snair, bottle.

Cone 10 (Derek Clarkson)

INGREDIENT	PARTS
Frit 3110	47.44
Zinc oxide	25.33
Silica	19.55
Titanium dioxide	6.43
China clay	0.75
Alumina	0.50
Total	100.00

Cone 10

INGREDIENT	PARTS
Pemco Frit P-283 (Fusion Frit F-413 can be substituted)	73.20
Zinc oxide	21.00
Silica	4.80
Bentonite	1.00
Total	100.00

Crystalline matt glazes

Cone 10 (John Tilton)

Apply thinly. Very runny. Needs colourants. Try copper, cobalt, tin, and pink stains. This glaze may also be fired in reduction and can be fired in any standard kiln firing.

INGREDIENT	PARTS
Nepheline syenite	42.50
Barium carbonate	32.50
Whiting	10.00
Ball clay	7.50
Rutile	7.50
Total	100.00

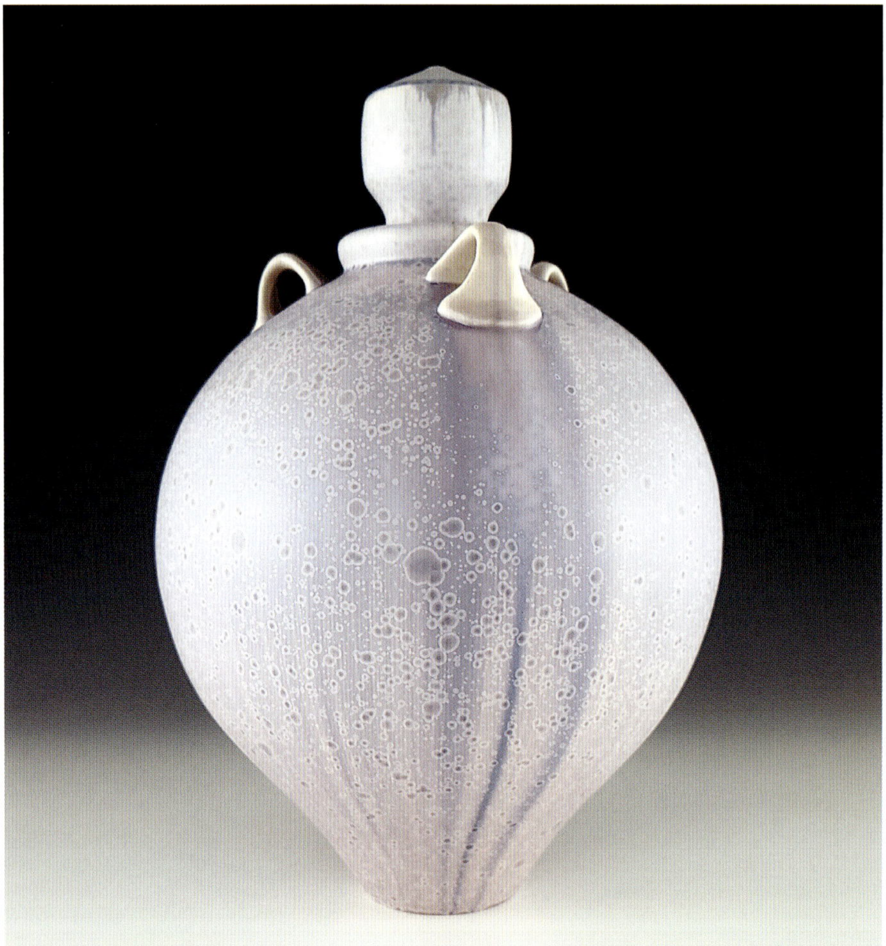

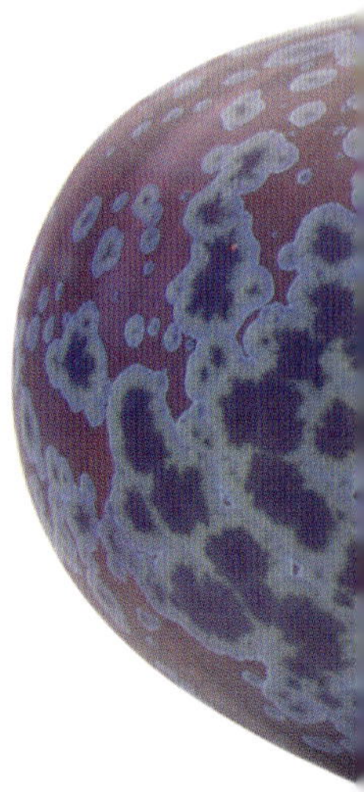

BELOW LEFT Crystalline matt © Ian Childers.

BELOW Crystalline matt © Andrew Krueger.

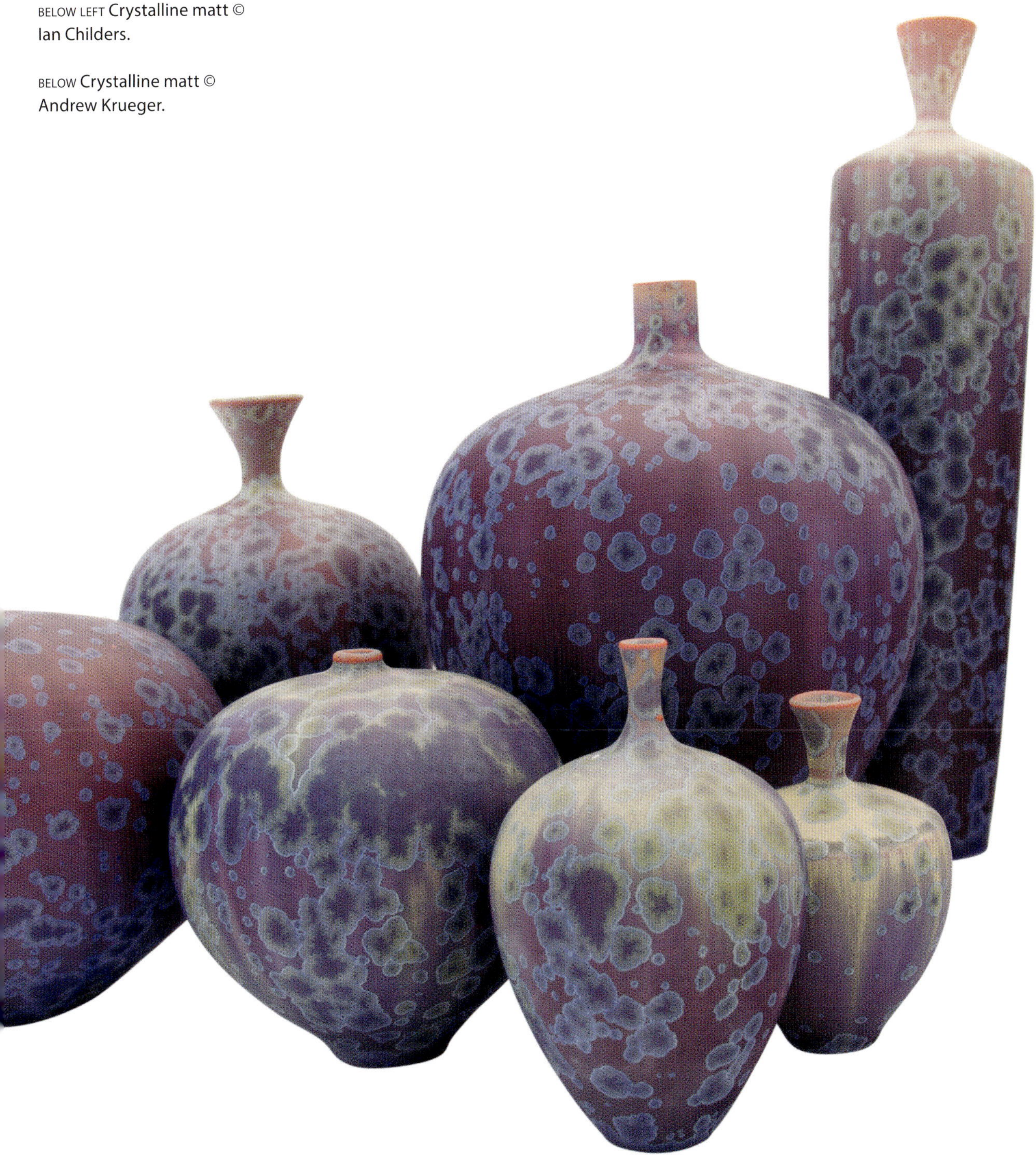

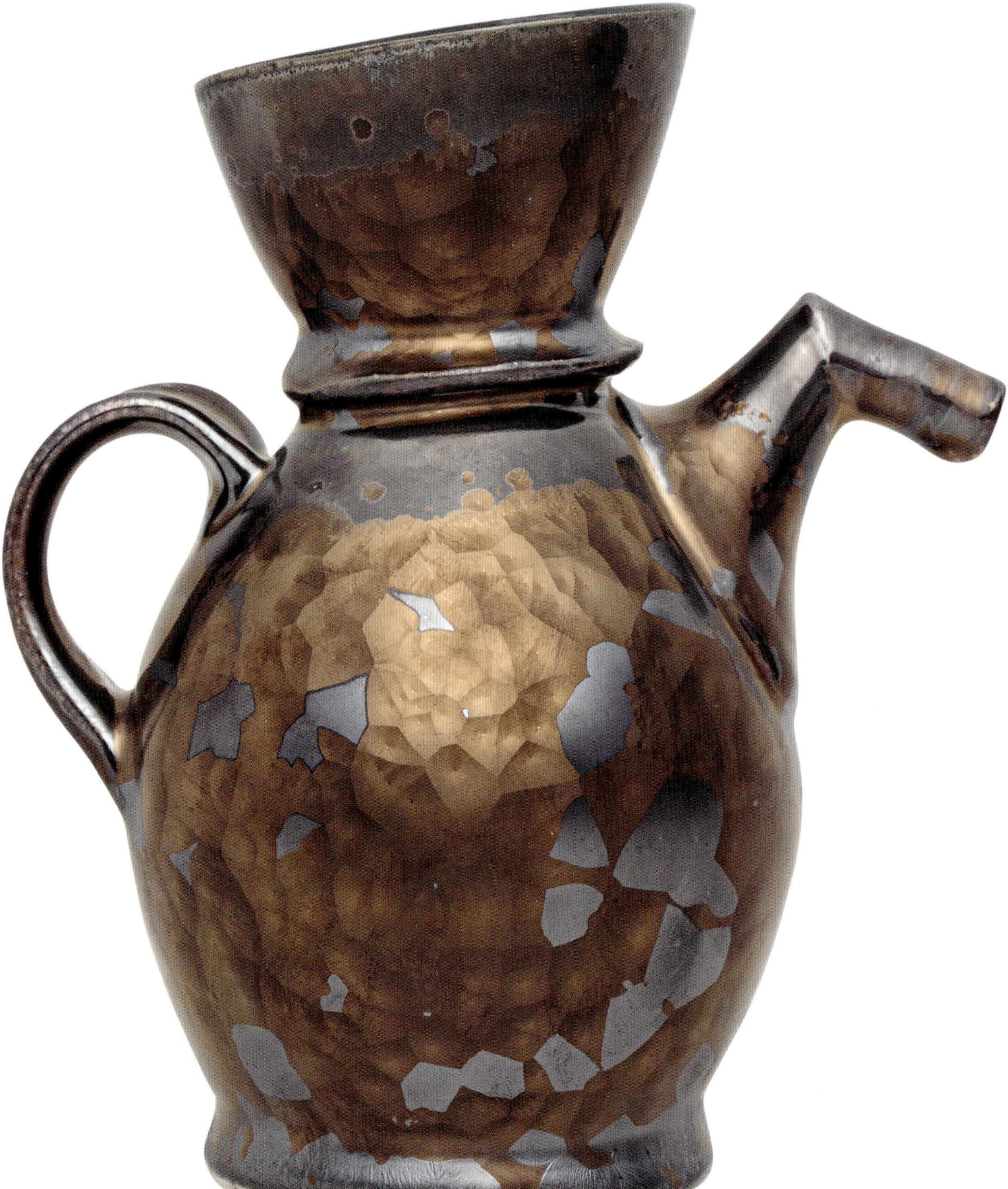

9 Aventurine glazes

ALL IMAGES Aventurine glaze © John Britt.

Aventurine glazes rely on a large percentage of metal deposits in the glaze to form very small shimmering crystals. Different elements and compounds can produce these crystals, and glazes can range in firing temperatures from cone 010 to cone 10. For our purposes, we will only be looking at the cone 9/10 aventurines, as lower-temperature glazes usually rely on lead as a flux.

Aventurines are fired to the maximum temperature suggested in the glaze recipe, and then the kiln is shut off. That's it. However, they do benefit from slow cooling and may even be fired with macrocrystalline glazes of the same temperature. Either method results in small but distinct crystals over the whole surface of the pot.

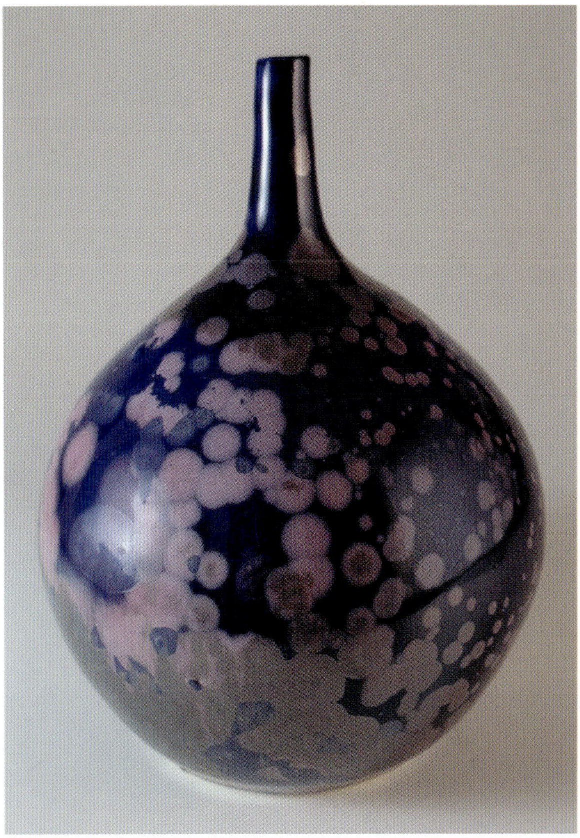

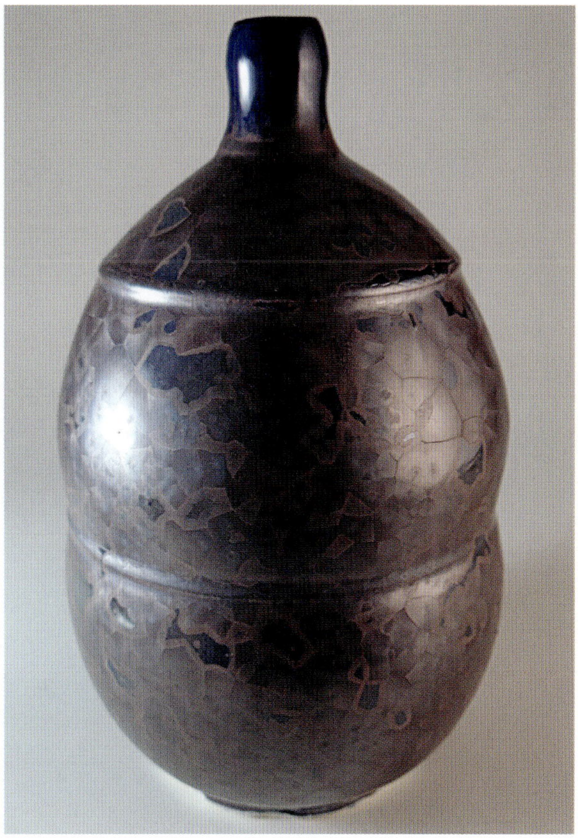

Ingredients in aventurine glazes

At lower temperatures, aventurines usually call for a frit. Those that contain unfritted materials sometimes have lead as a flux. For health and safety reasons, they should not be used in ceramics intended for serving food.

At higher temperatures, these glazes usually contain feldspar, whiting, china clay or kaolin (the alumina), and a frit. Boron can be added to increase fluidity but should only be used in small amounts or it may retard crystal formation. Silica also reduces fluidity.

Usually, iron is the principal oxide (10–20 per cent). Other colouring oxides can be used, including chromium, nickel, cobalt, and manganese oxides.

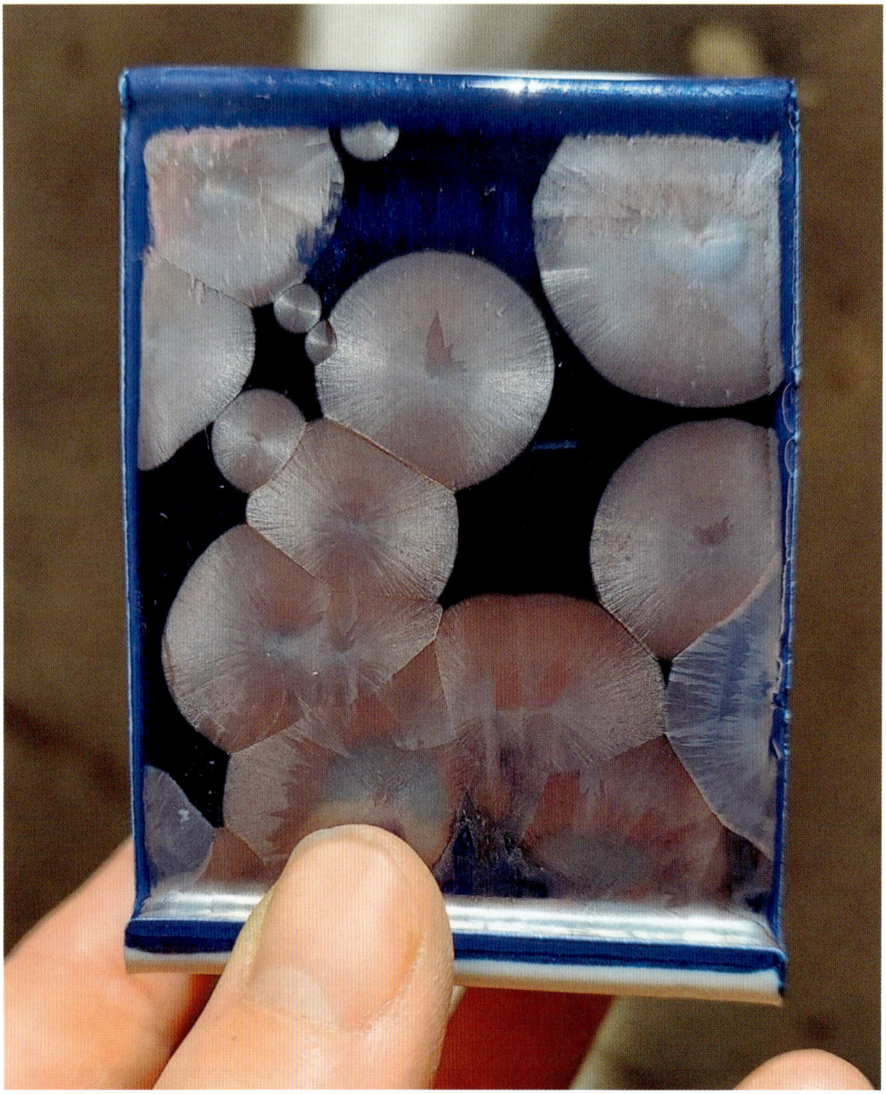

LEFT Close-up of aventurine glaze test tile with cobalt and copper © John Britt.

RIGHT Aventurine glaze on top of bottle, crystalline glaze below by Richard Bideau.

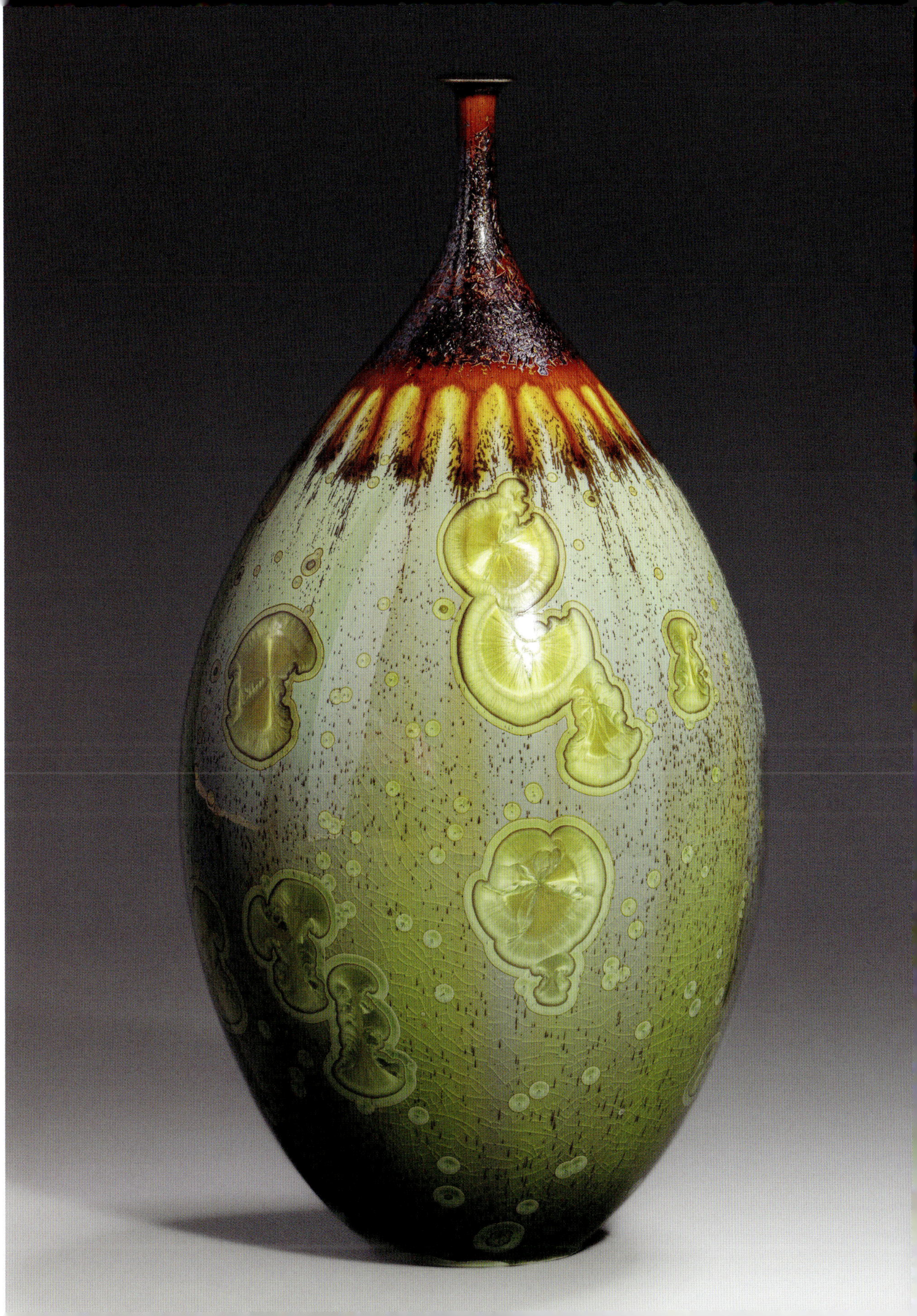

Richard Bideau (UK) fires his high-temperature aventurines alongside his zinc silicate glazes. In doing so, he has identified three distinct types of crystals in the iron aventurine, each associated with a different soak temperature and glaze formulation. By doing some extreme magnification photography, Richard was able to identify them.

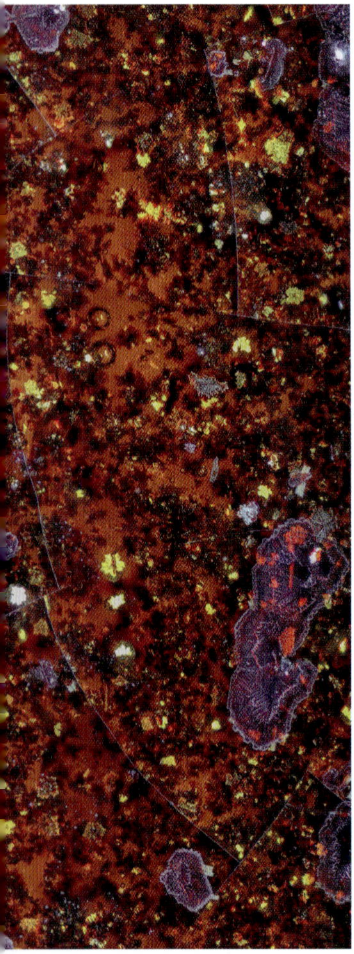

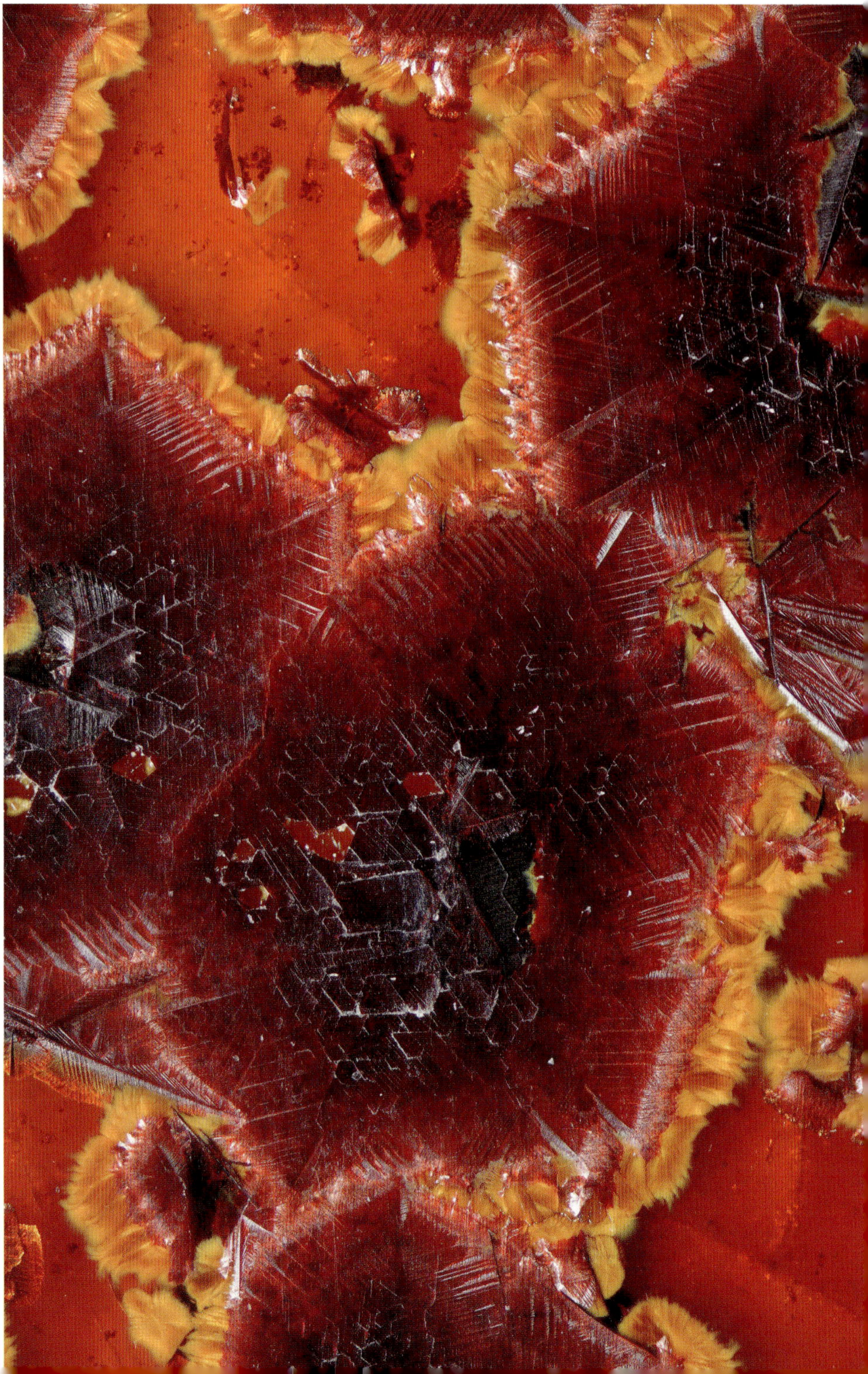

ABOVE AND RIGHT Aventurine glaze 5x magnified by Richard Bideau.

Glaze 9139 (p.68 far left image) contains two types of crystals that are almost certainly haematite; they are tiny and are responsible for the classic aventurine sparkle. One form grows as dendrites, something between a fern and a snowflake, while the other forms hexagonal plates. Which predominates depends on temperature and glaze composition. The crystals are oriented randomly and those that align with the light source reflect a golden sparkle.

Glaze 9139 (p.68 large image) contains a third type of crystal and is quite different; it forms on the surface and can grow to several millimetres in size. Richard suspects that glaze viscosity is the determining factor.

The chemistry of iron aventurines makes them react to changes in the glaze recipe in a way that is unlike most other glazes because of the opposed roles of iron's reaction in oxidation and reduction. In oxidation, red iron oxide acts in a similar way to alumina, stiffening the glaze; in reduction, it acts as a powerful flux. This is all subject to variations in the firing temperature and glaze composition, and leads to some interesting methods of varying the effect.

Iron aventurines are sensitive to maximum firing temperatures. Varying the amount of iron is one way to adjust to a particular firing cycle. A simple blend from 10–20 per cent of iron by weight of the total recipe will generally give the right amount. There will be a transition from runny, transparent amber glass (low iron) through to a dark, brown opaque glass with surface crystals (increased iron). In between, there should be a region that produces the desired effect. Richard cautions that the type of red iron oxide makes a difference; for the recipes to work correctly, it is important to choose the synthetic variety rather than the natural.

Aventurine: Cone 10 (Richard Bideau)

This is a stiff glaze that doesn't run and can be applied thickly.

INGREDIENT	PARTS
Frit 3110	45.50
Silica	25.30
Lithium carbonate	2.30
Calcium carbonate	2.60
Kaolin	9.70
Red iron oxide	14.70
Total	100.10

Aventurine: Cone 10 (Richard Bideau)

This glaze is runny, similar to zinc silicate macro crystals. Catch basins are recommended.

INGREDIENT	PARTS
Frit 3110	34.00
Silica	34.00
Lithium carbonate	3.10
Calcium carbonate	11.30
Red iron oxide	16.90
Total	99.30

Aventurine: Cone 10

INGREDIENT	PARTS
Potash feldspar	23.00
Silica	15.50
Colemanite	38.00
Red iron oxide	23.50
Total	100.00

Jamie Kozlowski also works with aventurines, but likes to apply the glaze over a red englobe containing 10 per cent dark red encapsulated Mason Stain #6088. Three to four coats of the englobe are sprayed on bone-dry porcelain, which is then bisque fired. After glazing, the pot is fired to cone 9/10, then the kiln is turned off and cooled. This glaze also works in a crystalline firing programme that has an extended hold.

Aventurine: Cone 10 (Jamie Kozlowski)

INGREDIENT	PARTS
Gillespie borate (a substitute for Gerstley borate – the mine is now depleted)	45.00
Custer feldspar	27.00
Silica	19.00
Lithium carbonate	9.00
Total	100.00
Add: Red iron oxide	28.00

Richard Bideau also has a chrome green aventurine. However, it must be fired on its own, separate from the zinc silicates, as the volatile chrome oxide is a powerful disrupter of other glazes.

Aventurine Chrome Green: Cone 10

This glaze runs. A catch basin is recommended.

INGREDIENT	PARTS
Frit 3110	41.00
Frit 3195	25.00
Silica	14.90
Calcium carbonate	1.20
Barium carbonate	11.90
Kaolin	6.00
Total	100.00
Add:	
Black copper oxide	1.50
Chromium oxide	1.50

BELOW Chrome aventurine glaze magnified x5 by Richard Bideau.

BELOW RIGHT Iron aventurine over red englobe © Jamie Kozlowski.

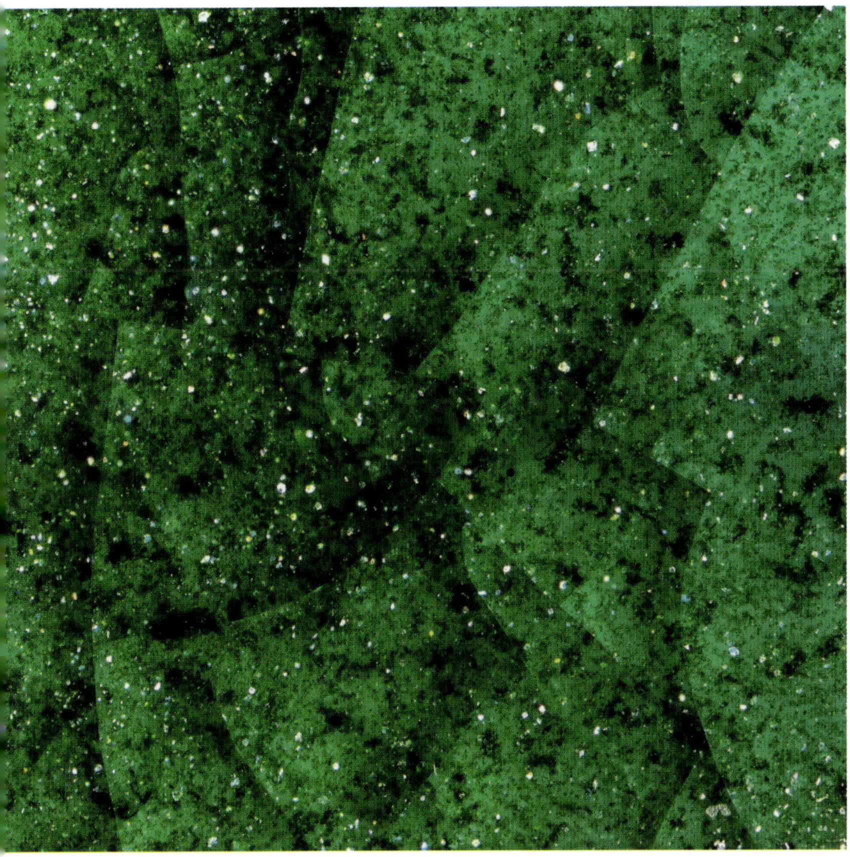

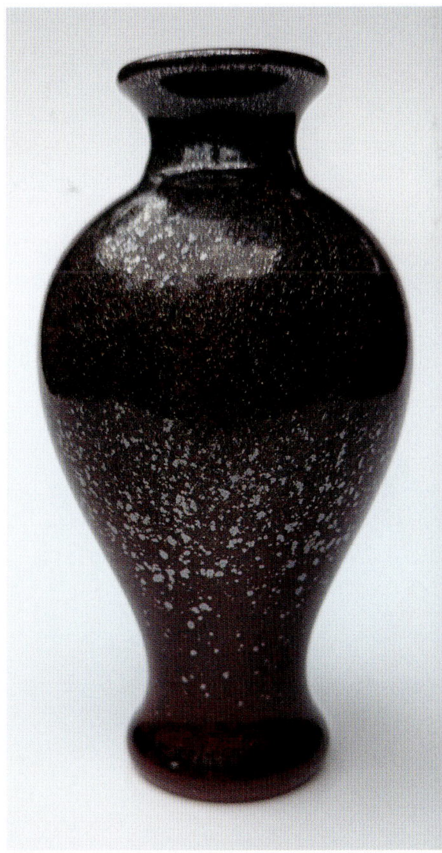

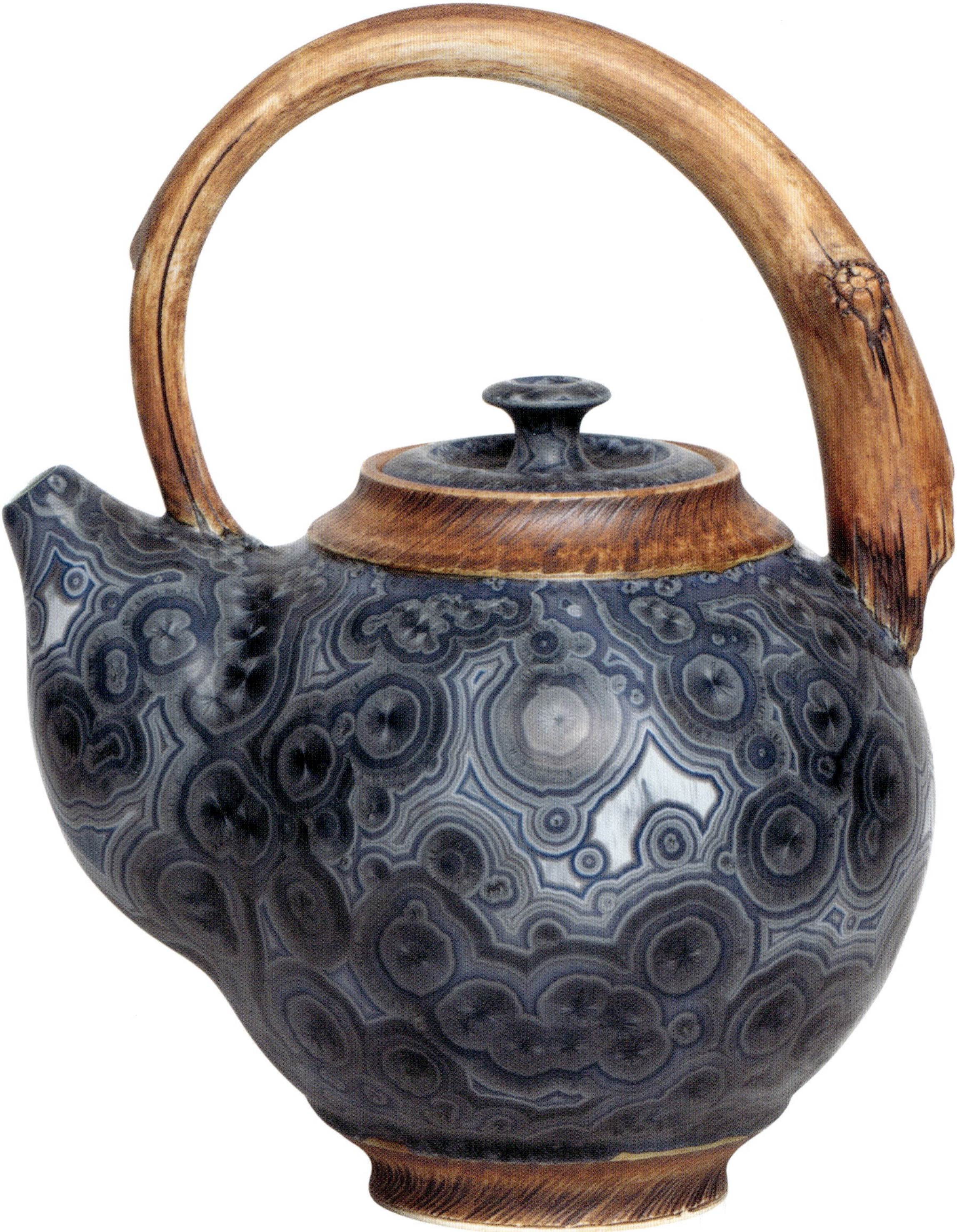

10

Glazing

LEFT Fired without a catch basin. Andy Boswell © Terry Boswell.

How we glaze is a personal choice. We each develop the approaches best suited to our surroundings, methods of working, and the materials available, but experimentation is necessary if an artist is to develop and grow. Careful notetaking and record keeping are essential.

Making the glaze

When making glazes, keep the work area clean and free from contaminants. Work in small batches of about 1,000g (35.3 oz) dry weight. Accurately weigh out ingredients using a triple beam balance or digital scale. Most come with a weighing pan and a balance that accounts for the weight of the pan. If your scale does not account for this, weigh the container that will hold the ingredient, then subtract its weight from the total.

After weighing all the ingredients, mix with a small amount of water to make a slurry. Pass this through an 80s screen twice to break up any coarse particles and to thoroughly mix. Adding a little water to the screen helps the ingredients through, but the glaze should be the consistency of heavy cream and applied thickly. The addition of a small amount of flocculant, such as Bentone® EW or magnesium sulphate (Epsom salts), retards settling of ingredients at the bottom of the container and allows you to stir the glaze more easily after it has settled. Approximately 0.25 per cent of the dry glaze weight dissolved in hot water is sufficient and has no apparent effect on the finish. The glazes need to be stirred well: for small amounts, an eggbeater works; for larger amounts, a paint mixer attached to an electric drill does a decent job.

Standard colour progression tests should be done for each individual glaze to determine the percentages most suitable for each colour additive. The following is a simple colour progression test.

For 1,000g (35.3 oz) dry glaze, place 100g (3.5 oz) of the colouring oxide on a piece of paper in a small pie-shaped pile. Use a knife or straight-edged tool and segment the pile, first in half, then in quarters. Take one quarter of the pile and halve it, then divide one of these halves into two equal parts. Mix the dry base glaze with water. Starting with the smallest section of oxide, add to the base mix and do a test. Add each of the segments in turn, making a test sample with each addition.

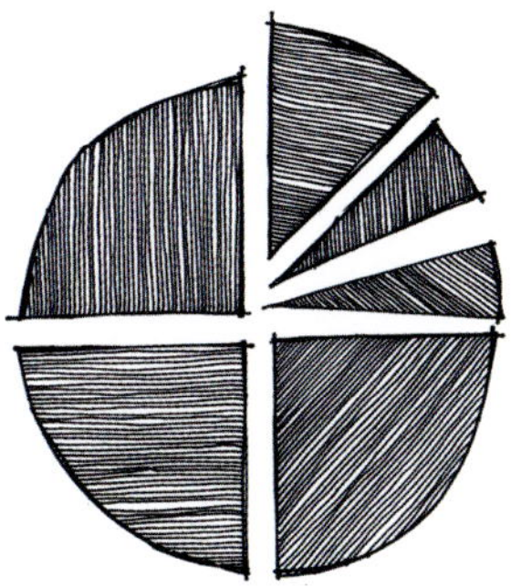

Applying the glaze

Before glazing begins, the bisque-fired pots should be cleaned with a damp sponge to remove dust. The glaze can be applied by dipping, brushing, pouring, or spraying. For smaller pieces, I use the attached pedestal as a handle, dip the pot into the glaze for three to four seconds, wait for about ten seconds, and then do a second dip only on the top half of the pot.

For larger pots, I use the one-piece catch basin and either brush, pour, or spray the glaze on. For brushing, place the pot on a banding wheel and, as the wheel slowly turns, apply the glaze in a thick coat with a soft, wide brush, applying a greater thickness at the top.

To spray, you must have a spray booth with adequate ventilation and wear a certified dust mask. As the pot turns on a wheel, apply the glaze with a greater concentration at the top.

Because the glaze tends to pool inside a bowl or plate, apply it more heavily around the rim, thinning towards the middle. A conventional glaze designed for the cone temperature can be used on the outside. Take care around the rim of the plate where the two glazes meet. Avoid direct contact or the crystalline glaze might run into the production glaze.

For items like vases and covered jars, a conventional glaze can be used on the insides.

Because crystalline glazes are ornamental, they should be reserved for decorative pieces. The roughness of some crystals, where they pool on the interior, makes them unsuitable for the inside of food containers; a production glaze should be used instead.

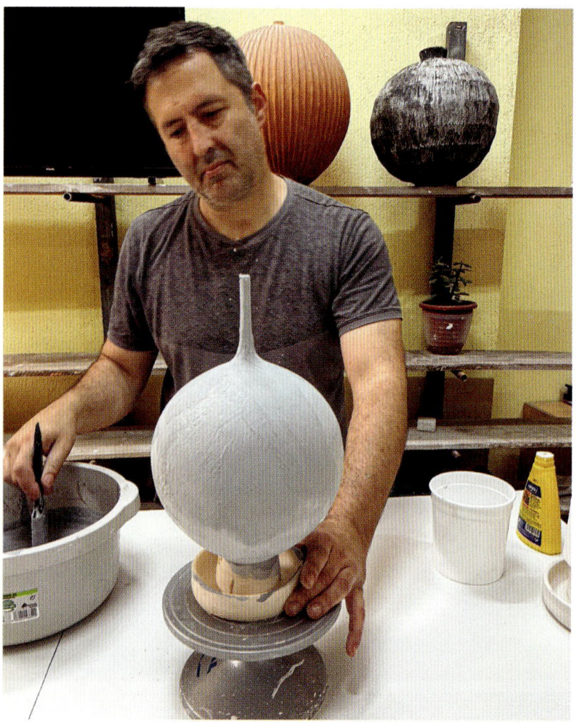

LEFT José Mariscal brushing on glaze © Jose MARISCAL Paneque.

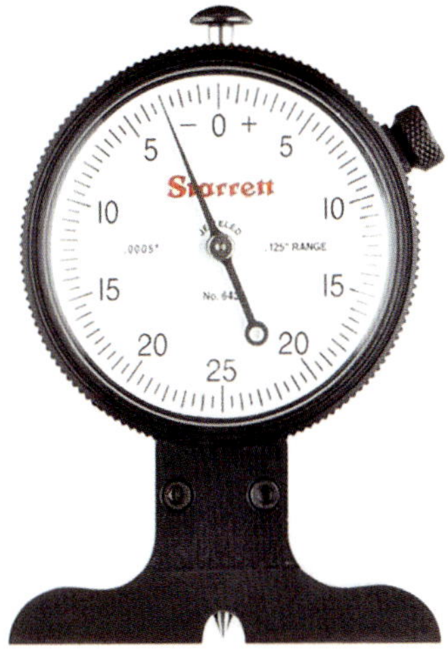

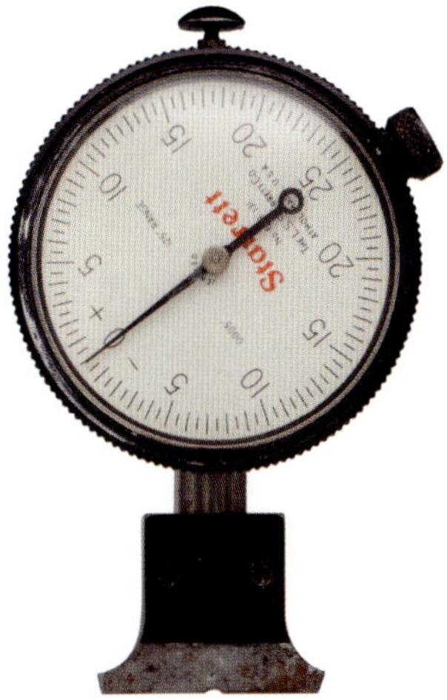

RIGHT John Tilton's depth gauge before modification. Both images © John Tilton.

FAR RIGHT Gauge after modification.

To gauge glaze thickness, John Tilton modified a Starrett 643JZ Dial Depth Gauge so that it could be used as a glaze micrometre. He rests the shoe of the instrument on the glaze surface, and the press of a button on the gauge causes a needle to extend through to the bisque pot, allowing him to take a reading. Through experimentation, John has learned how to apply the glaze to the ideal thickness.

When using insoluble materials, care must be taken as the glaze has no dry strength and brushes off easily. If applying glaze over a previously fired pot, adding gum tragacanth (the consistency of molasses or golden syrup) prevents it from running off during application.

Seeding

To 'seed' is to add nuclei to ensure crystallization occurs and, to some extent, placement of the crystals can sometimes be controlled by seeding. Ingredients that remain undissolved when the glaze becomes a liquid act as nuclei (seeds). They need not be the same ingredients as in the glaze recipe. Titanium dioxide, granular rutile, ilmenite, cadmium, molybdenum, tungsten, zinc oxide and zinc silicate may all be used as seeding agents. Experiments are necessary to determine how much seeding material to add to a specific glaze. Start by adding 0.5 per cent and then increase the additions by 0.5 per cent until you have the effect you want. Too great an addition will give a matt, rough appearance. Each seeding agent acts differently.

Alternatively, the pot is glazed with a crystalline glaze, minus seeding agents. The seeds are either pressed into or glued onto the dry glaze before firing. The fired pot should appear clear of crystals except in the specific areas where seed nuclei were applied. Seeding agents can also be glued onto the bisque-fired surface and a crystalline glaze, without seeding agents, is sprayed over the top.

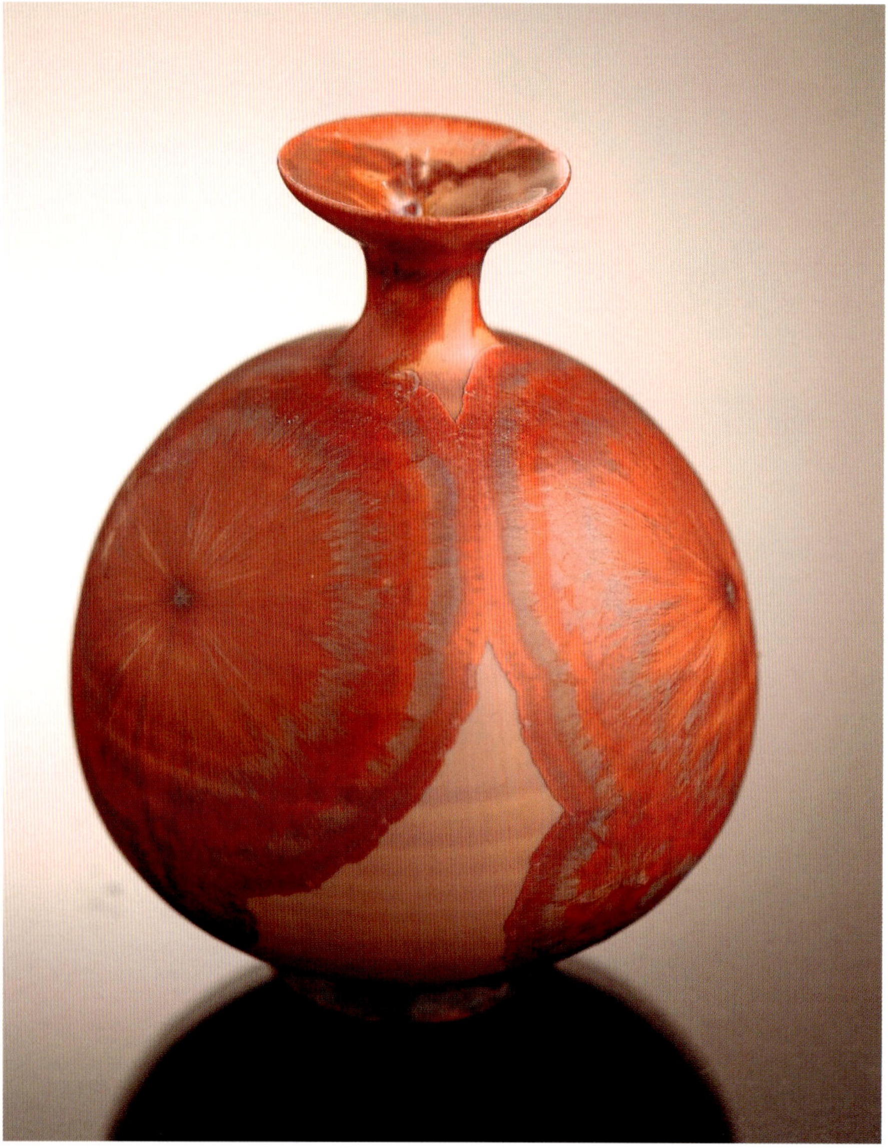

LEFT Seeded glaze.

RIGHT Close-up of molybdenum glaze © Holly McKeen.

Crystal enhancers

Crystal enhancers can change the crystal's shape, speed up formations, or increase growth. Some make crystals look more pronounced. Different ingredients give different results. Enhancers should only be used in tiny amounts, usually under 1 per cent.

Molybdenum, cadmium, and tungsten can change the crystal character, while iron and manganese affect the crystal's growth. Molybdenum is used in amounts of around 5 per cent. In smaller quantities, it promotes the formation of other types of crystals.

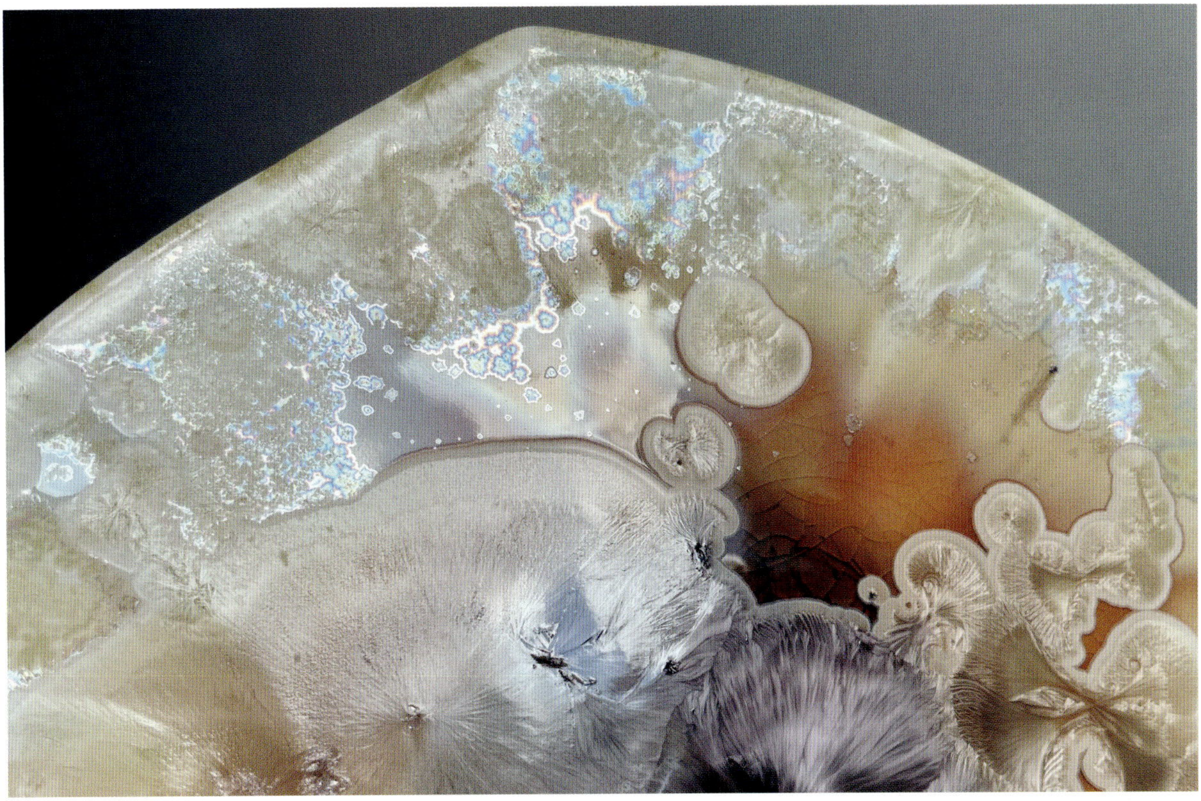

Glazing without a catch basin

Andrew Boswell (USA) has been making crystalline glazes without catch basins since 2018. It simply became a process of figuring out how much alumina to use, how thin the glaze coating should be, and how low he could fire the peak temperature to.

Andy's frit choice is Fusion Frit F-413 because it has a little more alumina than the more commonly used frit 3110. As an additional source of alumina, he incorporates Grolleg (a white firing, English kaolin) into the glaze in varying amounts, depending on how the recipe is performing. The additional alumina reduces running by stiffening the glaze. Andy has found, through experimenting, the ideal amount.

He is specific about having a target of 0.21g of dry glaze per square inch across the surface of his piece. If too thin, there isn't enough crystal-growing material; too thick and the glaze runs. He calculates the amount by multiplying the surface area by 0.21, which tells him how many grams of glaze to put on the piece. Assuming the glaze is equal parts materials and water, Andy doubles that number to account for the water weight. He simply weighs the bisque piece and begins adding glaze until it reaches the weight he desires. To achieve this amount, he combines the use of hydrometers, surface area calculations, brushing, and dipping.

The last variable is the firing temperature. It must be hot enough so that all the materials are fully melted, but anything past that point encourages running. Andy simply reduced the peak hold temperature by .56°C (1°F) for each consecutive firing until he got it right. That established a lower limit where he could achieve the melt he needed without going any further.

Andy's non-running glazes

Phil's Blue v2.2: Cone 9

INGREDIENT	PARTS
Fusion Frit F-413	59.30
Zinc oxide	20.50
Silica	12.20
Grolleg	1.00
Red copper oxide	4.20
Manganese dioxide	1.40
Cobalt carbonate	1.30
Red iron oxide	0.10
Total	100.00
Add:	
Veegum T (suspending agent)	1.50
CMC gum	2.00

Crystalline Jade v2.4: Cone 9

INGREDIENT	PARTS
Fusion Frit F-413	59.50
Zinc oxide	21.40
Silica	12.20
Grolleg	1.00
Red copper oxide	4.40
Manganese dioxide	1.40
Red iron oxide	0.10
Total	100.00
Add:	
Veegum T	1.50
CMC gum	2.00

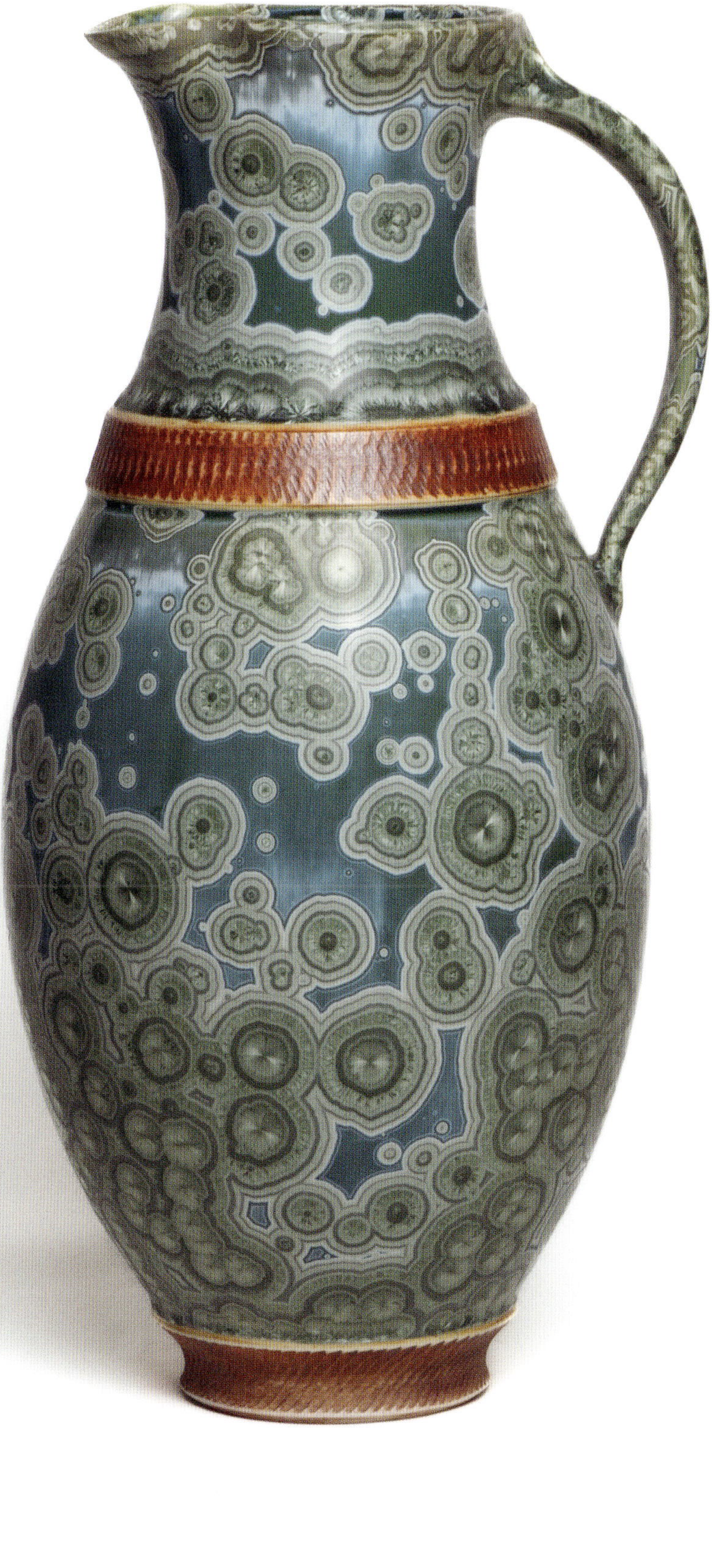

The results of Andy's tuning are crystalline glazes that don't run. However, this takes a lot of experimenting and should only be attempted by an advanced potter. If you do try this method, coating the kiln shelves with kiln wash or a covering of grog is a wise precaution when experimenting. Kilns vary and firings change as elements age, so what may work one time could give dissimilar results the next.

Another way to avoid using catch basins is to apply a matt glaze high in alumina to the lower portion of the pot. A crystalline glaze is applied, thicker at the top, thinning as it nears the matt section. Hopefully the matt glaze stops the flow, but protect your kiln shelves as a precaution. Where the crystalline glaze runs into the standard glaze, flow lines are created down the sides. The glazes chosen should be compatible in colour.

Making catch basins is extra work, but so is consistent testing. It all comes down to choice and individual preference.

LEFT Fired without a catch basin. Andy Boswell © Terry Boswell.

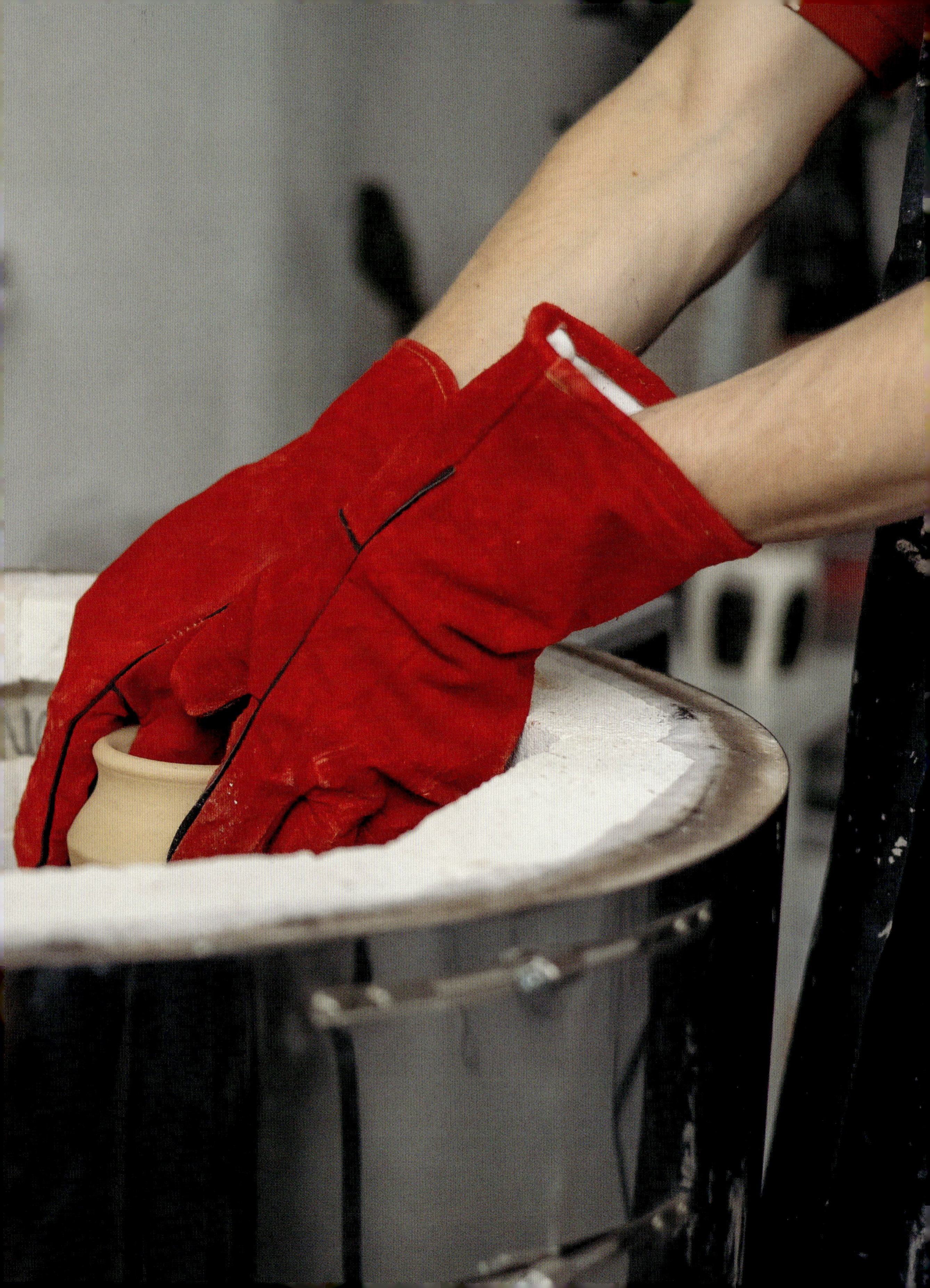

11 Kilns

LEFT Unloading a bisque kiln. Photo by Anna Reshetnikova via Getty Images.

The biggest equipment investment a potter will ever make is the kiln. Selecting one that can fire to the desired temperatures with ease, go through the various cycles in a crystalline firing, and maintain an even temperature, requires careful consideration. Since everyone's needs are different, a knowledgeable and reliable kiln distributor is vital.

The truism 'you get what you pay for' is particularly relevant when purchasing a kiln. Buy the best one you can afford. Know how many pots are going to be produced during a given time, the type of pottery being done, and your temperature requirements. Also know the amperage limit for your studio. Many potters who do crystalline glazes have a kiln for their non-crystalline studio work and use a separate, smaller one just for crystalline glazes. Having a fully automatic, programmable kiln controller is highly recommended.

Electric kilns

Macrocrystalline glazes, fired to cone 9 or above, require a kiln that can rise to this temperature easily. Going to these higher temperatures consistently wears the elements and thermocouple, and they degrade over time, taking longer to reach the desired cone. The same firing programme may give inconsistent results if the kiln is struggling to get to top temperature.

Most kilns made today are comparable, and are engineered to go to cone 10 or above using cone parameters. They are usually built with Kanthal® A-1 wire. Better yet, APM elements (elements designed to withstand frequent and extremely high temperatures, and holding ramps at high temperatures for prolonged periods of time), although more expensive, can withstand the demands placed on them by crystal firings to a much greater extent. If you are planning to do production crystal firings, choose a kiln with APM elements.

Kilns with a higher R factor help reduce fuel costs, provide a more even temperature throughout, and allow less heat to escape, making the kiln safer. A well-insulated kiln will also cool more slowly, usually giving nicer looking glazes; the elements also last longer because they don't have to work as hard. Because of the nature of crystalline glazes, having a small research kiln is helpful for doing glaze tests, but keep in mind that the results differ from kiln to kiln.

Electric kilns are either front- or top-loading. The front-loader is usually more expensive because of the bracing around the door, the reinforced door and hinge construction, and the flexible connections to door elements. Sometimes the door section may not contain elements, which could create a cool spot. The top-loader, although cheaper, may be more difficult to load, especially if it is a deep kiln. Bending down to load the kiln can cause back strain.

There are many types of kilns on the market. The most basic come only with variable heat dials. Others have bypassed the switches with a pre-set dial or a temperature controller built onto the kiln. This allows you to preprogramme a firing schedule. There are also programmable heat controllers that attach to and interface with the electrical supply.

Some kilns are more powerful than a crystalline firing requires, so it may be necessary to slow down the rate of heat rise to make sure the glaze has sufficient time to completely liquefy, and also to accurately control the rate per hour to a cone bend.

Gas-fired kilns

Fuel-burning kilns have a longer fully functioning ability and are capable of doing different types of firings. For crystalline firings they must be able to fire in both oxidation and reduction.

Several kiln manufacturers make gas-fired kilns suitable for crystalline glazes in a broad range of sizes and styles, including top-loading, chamber, car, and shuttle kilns.

Retailers specializing in kilns and ceramic equipment can recommend the most suitable kiln for a potter's needs.

RIGHT Geil gas-fired car kiln © Geil Kilns.

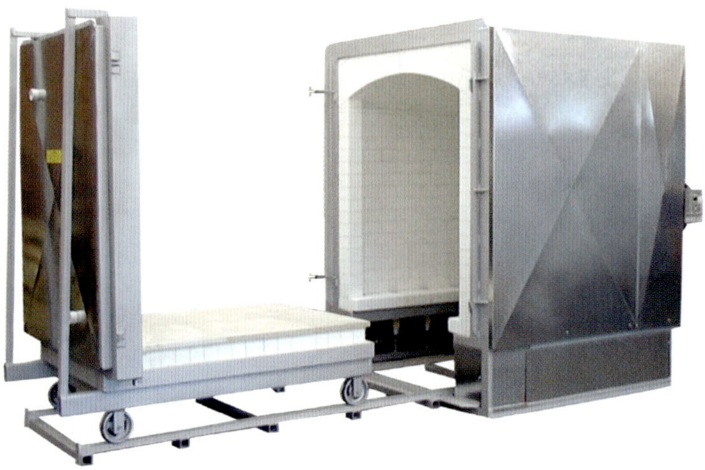

Rohde (Germany)

Rohde offers a large variety of kilns particularly suitable for crystalline firings. Their TE QT square top-loader, unlike most top-loaders, has heating elements mounted on ceramic support rods, rather than placed in grooves. This makes heating more efficient and the frequent change of elements easier. All Rohde kilns come with Kanthal® A-1 heating elements, and APM are available on special order

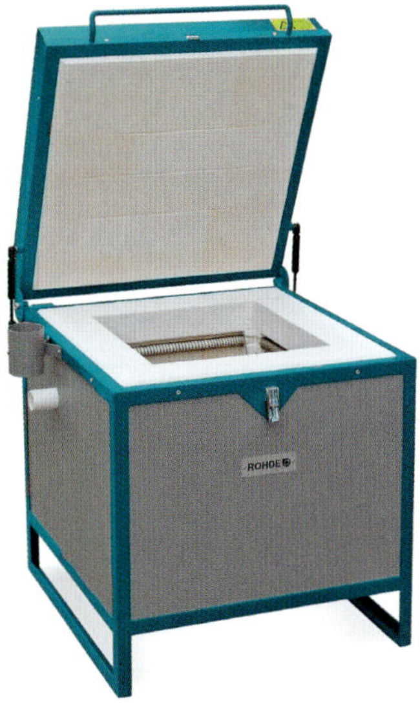

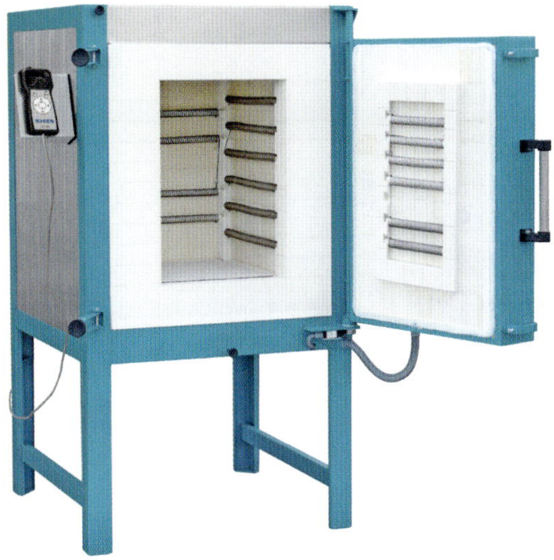

RIGHT Rohde top-loading kiln
© ROHDE Kilns.

FAR RIGHT Rohde chamber kiln
© ROHDE Kilns.

Rohde's chamber kilns also come with rod-supported heating elements and fire to cone 11. The LE-SH and KE-SH series are capable of 1400°C (2552°F), are well insulated with a robust brick lining, and are particularly suited to crystalline firings. They are available up to 480 L (17 cu ft) and come with APM heating elements.

L&L Kilns (USA)

L&L Kilns produce electric kilns ranging in size from 9.3 L (0.33 cu ft) to 2,832 L (100 cu ft), including top-loading and front-loading models. Their JH Series Crystalline kilns are rated for up to cone 12, are specifically designed for crystalline firings, and are recognised for reducing downtime through simplified maintenance and reliable performance. The company emphasises time savings, allowing users to focus more on their creative work rather than kiln upkeep. All L&L Kilns come with Bartlett controllers.

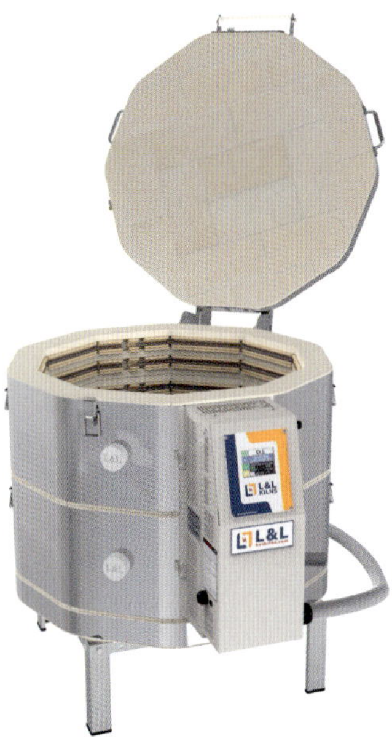

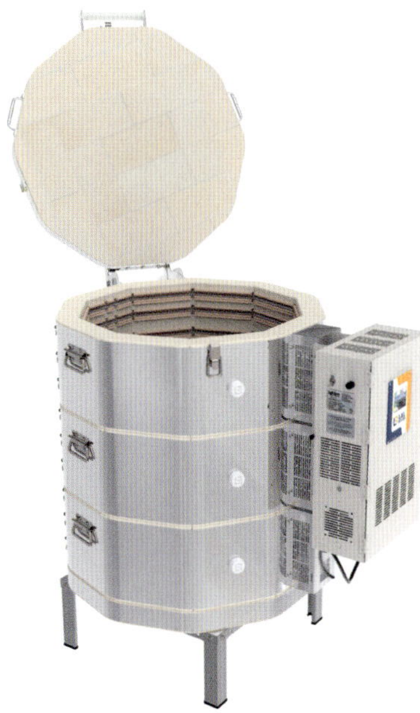

RIGHT L&L 2-ring top-loader
© L&L Kiln Mfg.

FAR RIGHT L&L 3-ring top-loader
© L&L Kiln Mfg.

Skutt Ceramic Products (USA)

Skutt Ceramic Products produces a large line of electric kilns. Their KMT-1027-3" has the power of a production kiln but comes in a smaller size, making it ideal for crystalline glazes. This kiln features 3-inch thick insulating brick, APM elements, and a Type S thermocouple designed for repeat firings to high temperatures. Skutt also makes several other models suitable for repeated high-temperature firings. The KMT-1018-3", KMT-822-3" and KMT-818-3" models all offer smaller firing chambers for smaller workloads, APM elements, and Type S thermocouple upgrades.

Geil Kilns (USA)

Geil Kilns manufactures electric and gas kilns. Their front-loaders are either brick- or fibre-lined and offer heavy-duty, welded, structural steel-frame construction, a built-in vent system, door safety interlock, and simple change elements, and all models easily fire to 1,304°C (2,380°F). The 10-gauge heating elements are coiled wire using Kanthal® A-1 alloy. The coils are held in grooved insulating fire brick using ceramic tube inserts. All their electric kilns feature Bartlett programmable temperature controllers which automatically power each zone as it ramps or soaks through the desired firing programme.

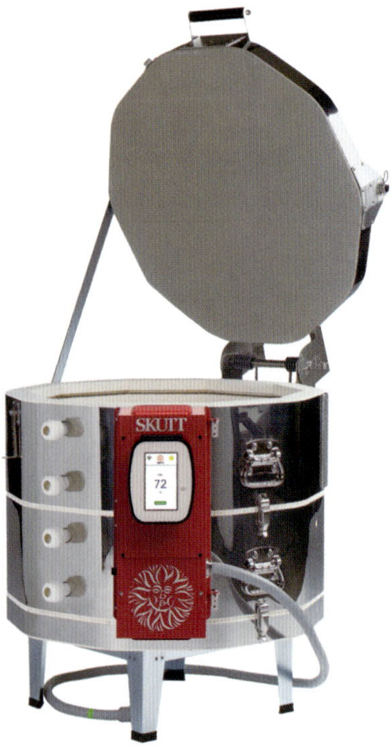

BELOW LEFT Skutt top-loader © Skutt Ceramic Products, Inc.

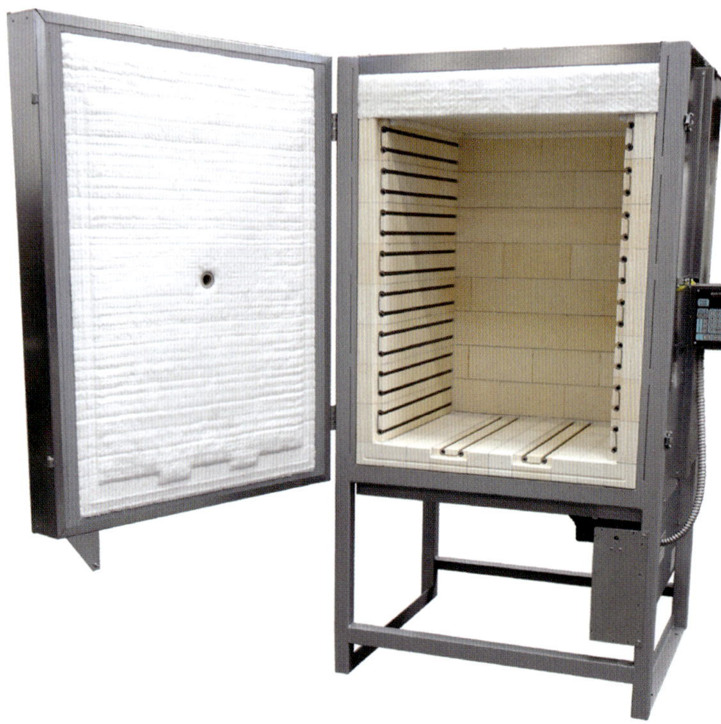

BELOW Geil chamber kiln © Geil Kilns.

Bailey Pottery Equipment (USA)

Bailey Pottery Equipment makes both gas and electric kilns with top-loading and front-loading models, and various shuttles (trackless and track style). The electric top-loaders offer significant extra insulation for retaining heat, the reduced heat loss saves energy and extends element life. The Bailey Thermal Logic Electric starts with a patented element holder system for retaining large size elements, offering more efficient heat radiation into the load; it also provides more power (BTUs/cu ft) to build heat at a faster ramp rate. Also, the lower range of watt loss density (per square inch) that is incorporated when calculating element output produces less stress on the Kanthal® wire and greatly extends their life cycle.

Tucker's Pottery Supplies (Canada)

Tucker's Pottery Supplies manufacture Cone Art kilns. Their electric kilns are all built to the same standards and are rated to fire evenly to cone 10. Every Cone Art kiln comes with extra insulation, a programme controller, multiple thermocouples, and floor elements, enabling their kilns to maintain an even temperature throughout.

Cone Art kilns take longer to cool down, partly because of the extra insulation; the other restricting factor is multiple thermocouples. Options provided are vent systems (which can either be independent or integrated into the kiln and operated by the controller), and a safety lid switch that cuts off power to the elements any time the lid is opened (standard on their CE[20] certified models, but may be optional in some local regions on CSA/cUL[21] certified models). They also offer the chance to upgrade to solid-state relays and Type S thermocouples. Cone Art kilns come with either the Bartlett Genesis 2.0 or V6-CF controller.

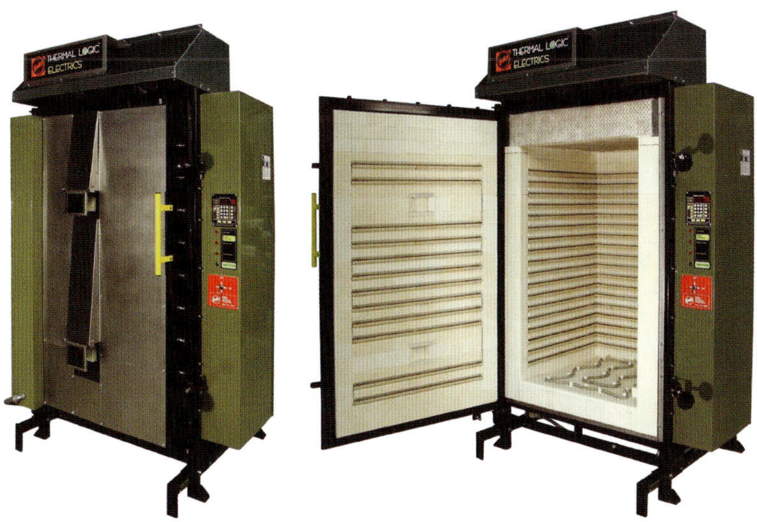

BELOW Bailey Thermal Logic Electric kiln, door closed and open © Bailey Pottery Equipment.

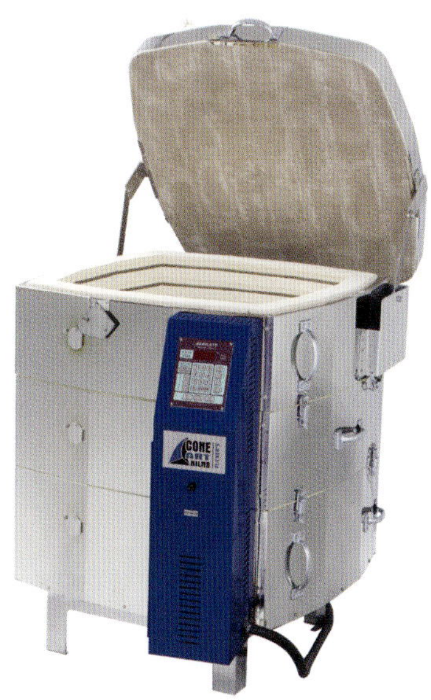

BELOW FAR RIGHT Cone Art kiln © Tucker's Pottery Supplies.

[20] CE certification indicates the models comply with European safety regulations.

[21] CSA (Canadian Standards Association) and UL (Underwriters Labaratories, regulating US safety standards) indicate that models comply with North American safety regulations.

Kiln preparation and safety precautions

Electric kilns

When you get a new kiln, read the handbook thoroughly. Placement in the studio is especially important. The floor must be level and made of non-combustible material such as concrete or brick. The kiln should sit at least 46 cm (18 in) from the nearest wall, and the wall should also be non-flammable.

It is important to have the kiln installed by a qualified electrician experienced in wiring kilns, to make sure it is properly grounded. Before the first firing, remove all packing material. Set the elements by firing the kiln empty, up to its maximum temperature; if you don't do this, the Kanthal® elements will contract from their recesses. The elements should also be oxidised as described in the owner's manual, as this extends their life.

- Keep the interior clean. Vacuum kiln and element grooves every few firings.
- Replace elements when they burn out or are no longer firing to full capacity. Long, drawn-out firings using weak elements can over-fire ware, seriously affecting the results.
- Change all the elements at the same time. This allows them to operate at equal intensity.
- Replace crumbling or broken bricks.
- When viewing the cones, protect eyes from heat and glare by wearing infrared protection goggles. Only infrared goggles shield from infrared radiation. Modern goggles feature feather-light polycarbonate lenses, which give nearly natural colour perception and a sharper image.
- If the kiln is not fitted with an exhaust fan, make sure the room is well ventilated. The exhaust fumes are toxic.
- As a rule of thumb, always be nearby to monitor the kiln.

Fuel kilns

Before even considering purchasing a fuel-burning kiln, consult your local authorities to check for any restrictions: safety requirements differ geographically. Your local fire department may need to approve or at least be made aware that you are considering a fuel-burning kiln. We were required to have an allowance of 1m (3 ft) all around our gas kiln, and the kiln-room walls had to be made of an inflammable material. Your kiln supplier will advise you on placement and set-up, but you will need a licensed gas technician to approve and connect the kiln before use. Don't forget to inform your insurance company.

Fuel-burning kilns are often placed in a separate building because they give off toxic fumes; even then, an extraction system must be in place. Allowances must also be made for a chimney to go through the roof or wall.

Loading the kiln

Since crystalline-glazed pots are fired in a collection dish, it is not necessary to kiln wash the shelves. A dusting of silica powder on the surface of the shelf prevents the catch basins from sticking. However, if there is a conventional glaze on the outside of the pot and running is a possibility, brushing or spraying the kiln shelves with kiln wash is highly recommended.

A traditional kiln wash for high-temperature firings is made up of 40 per cent alumina, 10 per cent silica and 50 per cent kaolin. These ingredients are mixed with water to a creamy consistency. Two or more coats should be applied with a wide brush, each coat applied before the previous one is completely dry.

- Use a three-point stacking system for support placement of shelves in the kiln. This reduces how much kiln furniture you use, and creates a more stable structure.
- Keep the loading uniform. Crowded areas take longer to heat, hold their heat for longer, can radiate heat to each other, and may produce a hot spot on the ware.
- Do not pack the kiln as tightly as you would for a normal firing.
- Make sure the pot is stable in the collection dish. A wobbly pot could topple, taking others with it.
- In a top-loading kiln, place a kiln shelf over the top, between the chamber and the lid. This protects ware from materials which may fall off the lid.

For all kiln firings keep a notebook on hand to record each firing and the results.

RIGHT Small electric test kiln. Photo by Vladimir Zapletin via Getty Images.

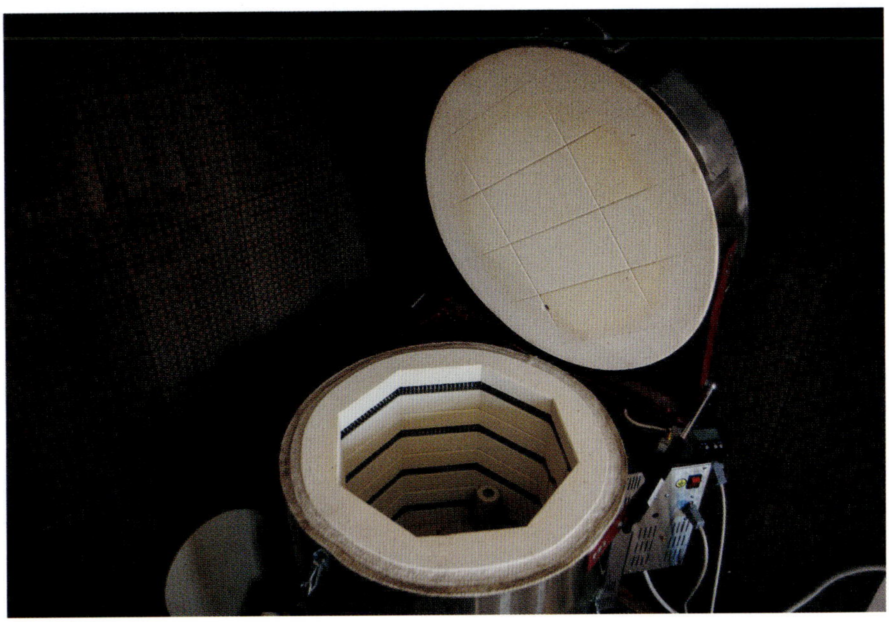

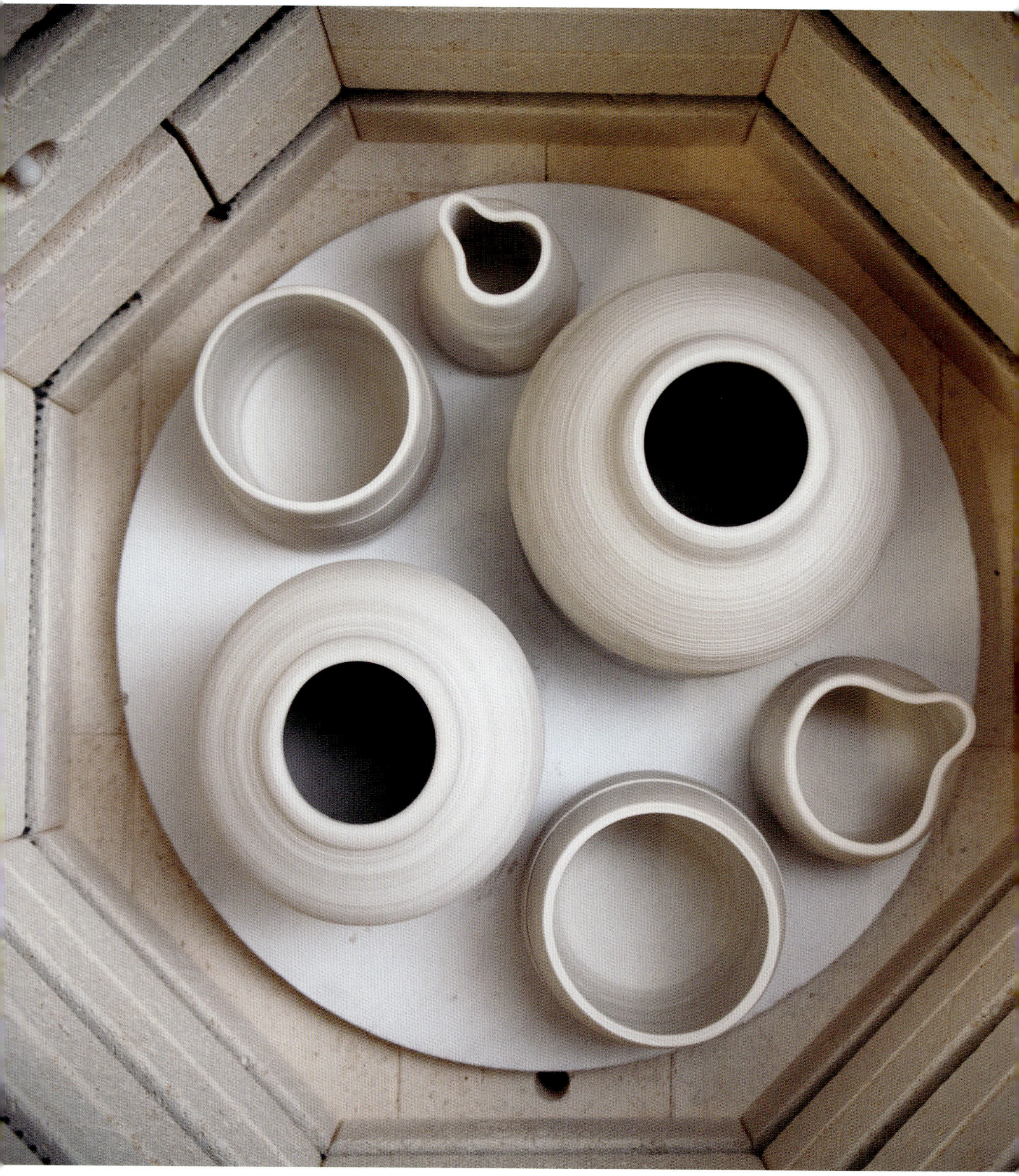

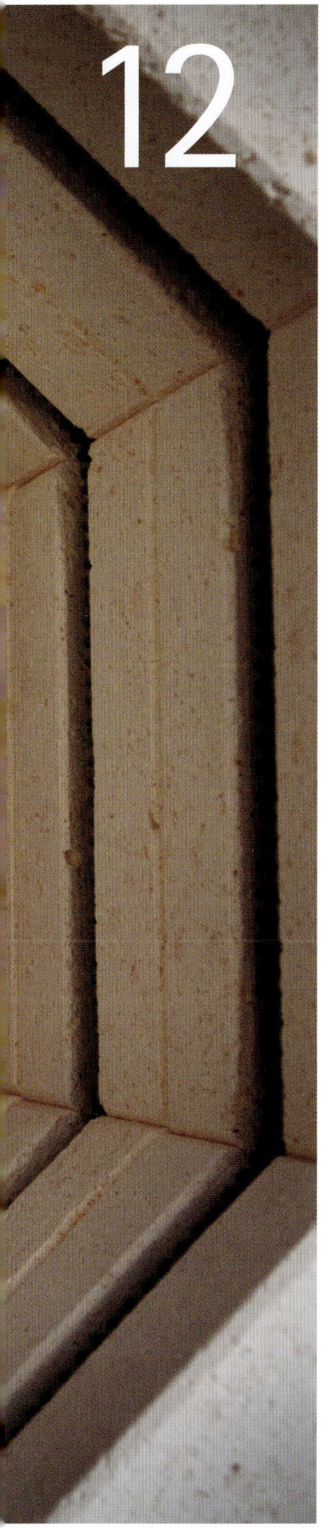

12 Firing aids

In the last chapter, I suggested buying the best kiln you can afford. The same advice applies to firing accessories. Usually, the better the equipment, the more consistent the results – especially important if you want to repeat certain firing patterns.

It is possible to produce crystals in a kiln that has variable heat dials using cones and a pyrometer. The results are often inconsistent, but with luck and understanding you may create some beautiful pots. If you are just starting to experiment with crystalline glazes and are using pre-established firing patterns, this basic equipment could be a starting point.

State-of-the-art apparatus does not guarantee satisfactory results; using it correctly, experimenting, and keeping records improves the chances of success. Starting with the barest essentials and moving to the most advanced equipment available, I will look at the firing aids on the market at the time of writing.

Pyrometric cones

Regardless of what firing instruments you may have, pyrometric cones are essential. They are made of ceramic material similar to a glaze, and react by melting and bending when exposed to a certain temperature over a certain length of time (heat work). Cones also indicate heat distribution in your kiln.

Cones are numbered on a scale, where the midpoint is 1, increasing to 2, 3, 4, etc., up to 42, and decreasing 01, 02, 03, etc., down to 022. Each cone number relates to a designated amount of heat work done, as influenced by time and temperature parameters. For example, Orton standard pyrometric cone 9 deforms at approximately 1280°C (2336°F) given a heat rise of 150°C (270°F) an hour.

Usually, cones are used in groups of three. They are placed in a cone pack which holds them at an 8° angle from vertical. The first is the guide cone. When it starts to bend, the firing is close to maturity. The second is the firing cone, and this tells you that the desired heat work has been achieved. The third, the guard cone, indicates over-firing. A cone temperature is reached when the firing cone has deformed so the tip is even with the top of its base.

Cone packs are placed in the kiln to be viewed through the spy holes; make sure they are not placed too close to the elements. At red heat and above, they become visible. They can be used only once. Until you know your kiln well, there should be a cone pack on every shelf.

Cones monitor heat work based on temperature rise and indicate the temperature of the pots rather than the air temperature, which is usually hotter. To measure rise and fall and soaking temperatures, potters must rely on other devices that indicate air temperature only.

PREVIOUS PAGE Ready for firing. Photo by tagphoto via Getty Images.

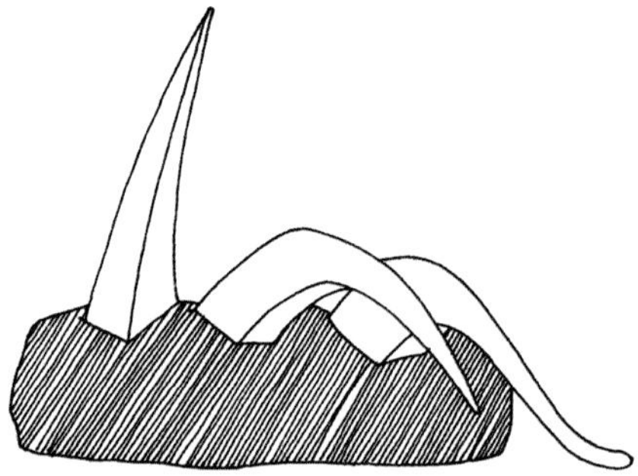

LEFT Pyrometric cones measure heat work done, which is the combined effect of temperature and time during a kiln firing.

Probes/thermocouples

The thermocouple (or temperature sensor) is a probe which is inserted near the midpoint of the kiln, several inches from the wall. It is made of two dissimilar wires fused together at one end. When the ends are heated, an electromotive force is generated, and a voltage develops across the wires. This voltage is roughly proportional to the kiln temperature. Analogue (needle type) or digital thermocouples electronically convert the voltage measurement to the corresponding kiln temperature and then display the temperature directly.

Most modern standard kilns come equipped with controllers and standard Type K thermocouples. The thermocouple and controller are calibrated to work together. For firings of cone 9 or less, a Type K (chromel/alumel) thermocouple may be used. The chromel (positive) wire, is an alloy of approximately 90 per cent nickel and 10 per cent chromium; the alumel (negative) wire is an alloy of approximately 94 per cent nickel, 2 per cent aluminium, 3 percent manganese, and 1 per cent silicon. This type of thermocouple is relatively inexpensive, but it's accuracy declines with use. The fusion points tend to oxidise and deteriorate rapidly at higher temperatures. It is not unusual to have an error of more than 14°C (25°F) when fired repeatedly to cone 6. Cost effectiveness however, favours using a Type K thermocouple, replacing it fairly often, and realizing it is only a temperature guide, not an accurate measurement.

Thermocouples can also be affected by atmosphere. Some materials in the kiln react with the thermocouple wire, changing its composition and resulting in temperature measurement errors. Protection tubes can solve the problems caused by atmosphere, but they can delay the thermocouple's response to heat adjustments, creating a difference between actual and reported temperatures. The tubes also have a wicking potential, which may give a false temperature reading of up to 10°C (18°F) less than the actual temperature.[22]

[22] Vukovich, Milan Jr and Dale A Fronk, 'The Role of Pyrometric Cones and Temperature in the Firing Process', in *Ceramic Manufacturing Council – Kilns and Firing*, ed. John B Wachtman, The American Ceramic Society Inc., Columbus, Ohio, 1990, pp.1905–1921.

For high-temperature or high-precision firings (cone 9 or above), a Type S or Type R thermocouple is highly recommended. The Type R is more common in kilns built in the United Kingdom and Australia, the Type S in kilns manufactured in continental Europe and North America. Both are made of a positive wire containing platinum and rhodium, and a negative wire of pure platinum. These thermocouples have a high resistance to oxidation and will indicate temperatures accurately up to about 1480°C (2700°F). Even though they are more expensive, for high-temperature and high-precision firings such as crystalline glaze work, the Type S and Type R thermocouples are more accurate and longer lasting.

It is necessary to use a thermocouple-based temperature measuring system after initial cone temperature is reached, and throughout the rest of the crystalline firing. If you keep accurate records and learn what temperatures correspond to measurements taken, you can make adjustments for discrepancies.

WARNING: It is important that the controller is set to the correct thermocouple type. Using a thermocouple set for the wrong controller will cause serious over-fire or under-fire. The extension wires must also be used on the correct type of thermocouple. If changing a thermocouple type, be sure to change the extension wires. Make sure the software and jumper settings match the type of thermocouple and extension wires you are using.

Pyrometers

A pyrometer is a calibrating instrument attached by electrical wire to a thermocouple. Pyrometer systems are calibrated to indicate temperatures reached during heating and cooling. They are reasonably priced, practical instruments for use with bisque and glaze firings that do not require the degree of accuracy needed for crystalline glazes. Most increment scales on pyrometers are small, and gradual changes in temperature are barely noticeable. Indeed, significant temperature changes can occur before the potter reacts and makes necessary adjustments. Therefore, for crystalline glaze firings, a more accurate digital thermometer is recommended.

Digital thermometers

Digital thermometers are temperature indicating devices with digital displays. They attach to a thermocouple and can convert the voltage output to a read-out in either Celsius or Fahrenheit. Differences are indicated to within a percentage of a degree, so even the slightest change in temperature is immediately noticeable.

These devices are designed for specific thermocouple junctions, so it is important that the read-out be matched with the thermocouple wire being used. Thermocouple sensors can be placed throughout the kiln, and because digital thermometers can be handheld, they can be plugged into and unplugged from various thermocouples, allowing comparison of different areas of the kiln. A digital thermometer gives a faster response to temperature fluctuations, allowing the potter to react and adjust the temperature in the kiln more quickly than when working with a pyrometer.

Digital programmable controllers

A digital programmable heat controller is highly recommended, especially for achieving repeat firing patterns. It means that you don't need to monitor the kiln for hours at a time, and is therefore a worthwhile investment.

Most modern standard kilns come equipped with controllers and Type K thermocouples; sometimes the controller can be purchased separately and installed by an electrician. The controller must interface between the power source and the kiln's main switching panel.

A controller comes with ramp and hold features, and can be programmed to regulate electrical input by a number of ramps, to allow for various soaks and slow cooling cycles, and to shut the kiln off automatically. Some can fire more than one kiln at a time, each with its own firing schedule. Controllers also allow the operator to duplicate and repeat firing patterns, consistently following the same profile as long as the kiln elements and the thermocouple have not aged too much. It will report firing times and top temperatures achieved, so any changes in kiln performance can be monitored and adjusted. Controllers allow the potter to check vital statistics during the firing, such as the rate of temperature rise and fall, firing progress, and temperature. Once it has been programmed and turned on, the operator doesn't need to make additional adjustments during the firing.

Many controllers allow you to programme by cone number, construct your own profile, or combine the two approaches. Most crystalline glaze firings consist of four to five segments on the way up, and three to five segments for cooling. Each involves a ramp to a temperature, and, when cooling, a hold at that temperature.

Controllers vary from simple to complex. Rohde kilns come with ST 310 controllers and Type S thermocouples as standard, but the optional ST 411 is better suited to crystalline glazes. The ST 411 is a single-zone controller and allows 32 programmes with 32 segments. Each segment consists of ramp gradient, target temperature and dwelling time. This controller can easily be exchanged between kilns, and can be handheld or wall mounted.

For chamber kilns of more than 250 litres (9 cu ft), a three-zone controller is recommended. All controllers come with a USB interface for data-logging on a USB-thumb drive. Firing data, such as actual setpoint and actual value for every minute, can be visualised by using the free Microsoft Excel tool ROHDEgraph, an offline tool for archiving firings and storing data locally. All Rohde ST controllers are compatible with their free Rohde myKiln app – the digital firing logbook.

The app enables the potter to:

- Check the live status of the kiln remotely at any time.
- Get push notifications for kiln status (i.e., target temperature reached), changes, or faults.
- Comfortably edit new firing curves on your mobile device or desktop computer, for use with different kilns.
- Get an instant visualization of a firing curve for your new programme.
- Get an automatically generated firing curve (setpoints and actual values) for every firing.
- Archive firing curves, and save own images and notes.

This searchable archive is a particularly useful tool because firing data, notes, and photos are available and accessible to the account holder. Data are securely and safely stored on the free app data server.

Bartlett Instrument Company (USA) is a major electronic controller manufacturer supplying controller instruments worldwide. All Bartlett controllers have error codes for troubleshooting and kiln maintenance, pre-programmed firing schedules (they store several programmes so users can enter and save custom programmes), the ability to add offsets, PID control algorithms (proportional–integral–derivative, meaning that the algorithm includes feedback and adjusts based on what has happened previously), and different speeds for firings. All touchscreen controllers have graphing of a programme and Wi-Fi connectivity so users can monitor their kiln from the KilnAid app. The Genesis 2.0 and Genesis Mini both have a Novice Mode that helps guide new kiln users. The Genesis Spark's new user interface comes with a history option so users can review and restart a recent firing (up to 10 firings).

Bartlett's V6-CF model previously featured remote programming through K.I.S.S. (Kiln Interface Software System). However, K.I.S.S. is no longer actively supported or sold. Bartlett now supports use of KilnAid, which can supervise the kiln firing, keeping both a visual and numerical record (looking at a graph often tells the potter more about what the kiln is doing than looking at numbers). KilnAid firing programmes can be archived and reused, and the historical record gives a better overall view on how the kiln has been performing. When the potter is away from the studio, they can communicate with the kiln via computer, tablet, or phone.

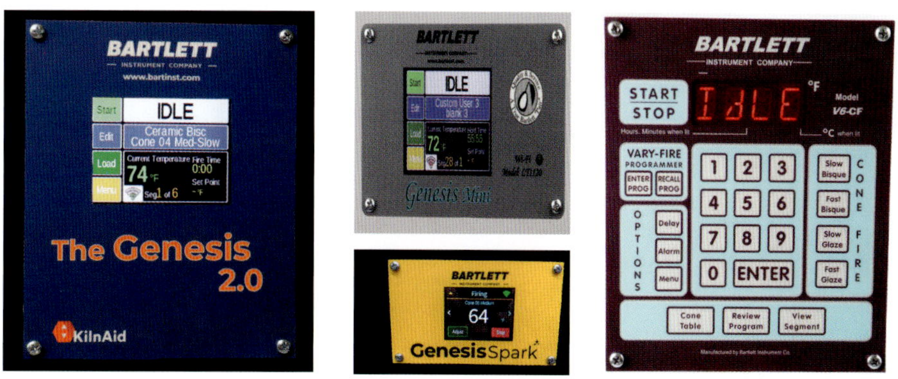

RIGHT Examples of digital programme controllers © Bartlett Instruments.

Skutt's KilnMaster Touchscreen (KMT) controller allows users to store up to 36 custom programmes, each with up to 16 individual segments. The large screen allows you to view the entire programme at one time, which makes confirming details much easier than a standard keypad controller. Custom programmes on a KMT can also be edited during a firing, and segments can be skipped mid-firing if desired. The controller is also Wi-Fi enabled and can be linked to the free KilnLink app for remote temperature monitoring. A premium KilnLink subscription can be purchased, allowing firing details to be logged for better historical record keeping and maintenance, and which allows the user to receive text or email alerts for errors or when the kiln has reached a desired temperature.

Stafford Instruments is a small UK-based manufacturer of kiln temperature controllers, pyrometers, temperature trips, and thermocouples suitable for pottery and glass firing kilns. All their controllers are designed to be easy to use and reliable, and incorporate extensive safety features. Controllers are manufactured in house, and the company is able to offer prompt expert technical support. Their ST215C controller is suitable for any single-zone kiln equipped with a controller socket. It offers 32 programmes, each with up to 32 ramp/soak segments. It can be supplied with an optional Wi-Fi interface that allows it to connect to their server via any domestic or commercial Wi-Fi router.

By logging into their Kiln Portal website, the potter can view the status of their kiln firing. The Kiln Portal can be configured to automatically log all firings, capturing the time and date of the firing, details of the programme used, and a record of the kiln temperature history. The Kiln Portal can also be used to upload, view, edit, and download programmes. The Wi-Fi also comes with a USB interface which allows logging of kiln temperatures to a USB flash (thumb) drive. The ST215C comes fitted with a plug lead and a wall mounting bracket. Stafford also offers several other design options.

One important feature of being able to track the progress of a firing is the ability to compare the programmed temperature setpoint with the actual temperature on a graph. If the elements are ageing and no longer keeping up with the programmed temperature rate of climb, it is clearly visible on the graph and the potter knows it is time to replace the elements, or to design a new, slower firing programme that works with crystalline glazes. The availability of these processors means that firing crystalline glazes is now much less complicated, which has opened the door for many more potters to work with them.

Note: Some potters set their controllers for a specific temperature rise or drop. Others use ramp up or ramp down and have the kiln temperature rise or drop at the rate the kiln is capable of.

European controllers use ramp rates directly, either in °C/hour or °F/hour. For the maximum heating and cooling rates, they use 'FULL' or 'SKIP' (depending on the manufacturer). Because of the differences in how controllers are programmed, I will only be using the terms 'Ramp up' or 'Ramp down' throughout the book when describing a rapid climb or drop in temperature.

Most American controllers can be programmed by a code to do ramps. These codes differ by the manufacturer, depend on whether you are firing in Celsius or Fahrenheit. Consult your controller handbook to determine what code to use for ramping up or down at the maximum rate.

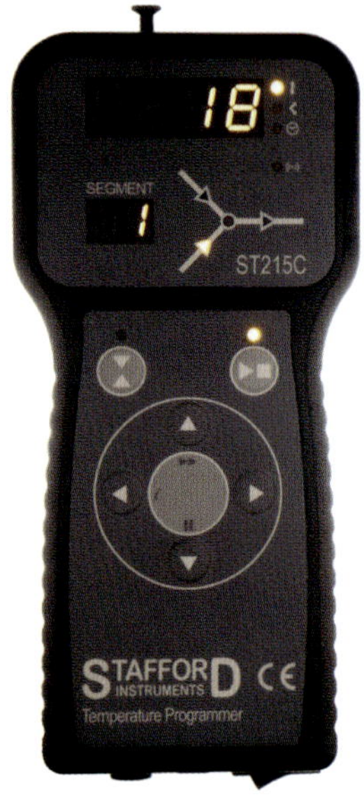

ABOVE Stafford controller © Stafford Instruments Limited.

RIGHT Loading a glaze kiln. Photo by Maskot via Getty Images.

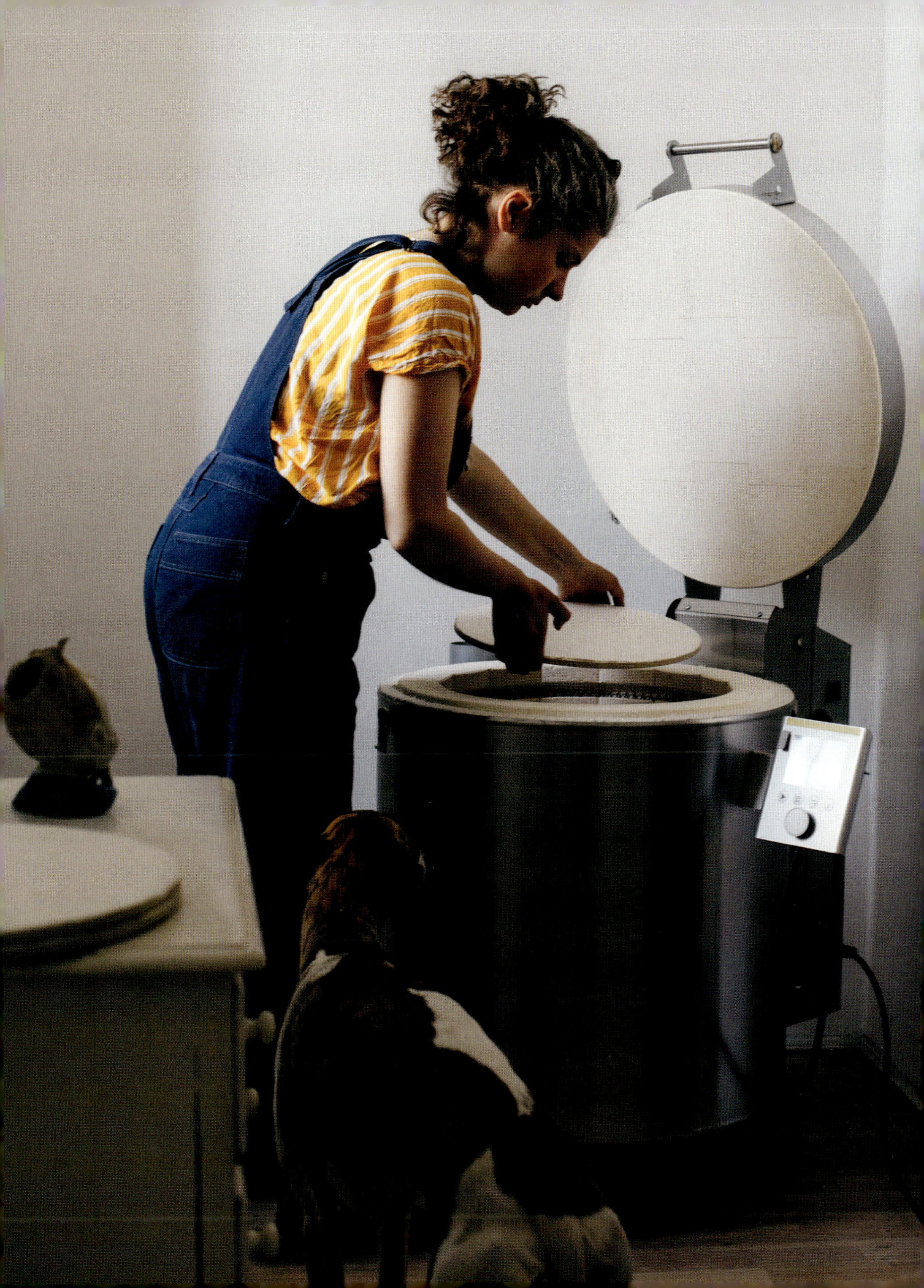

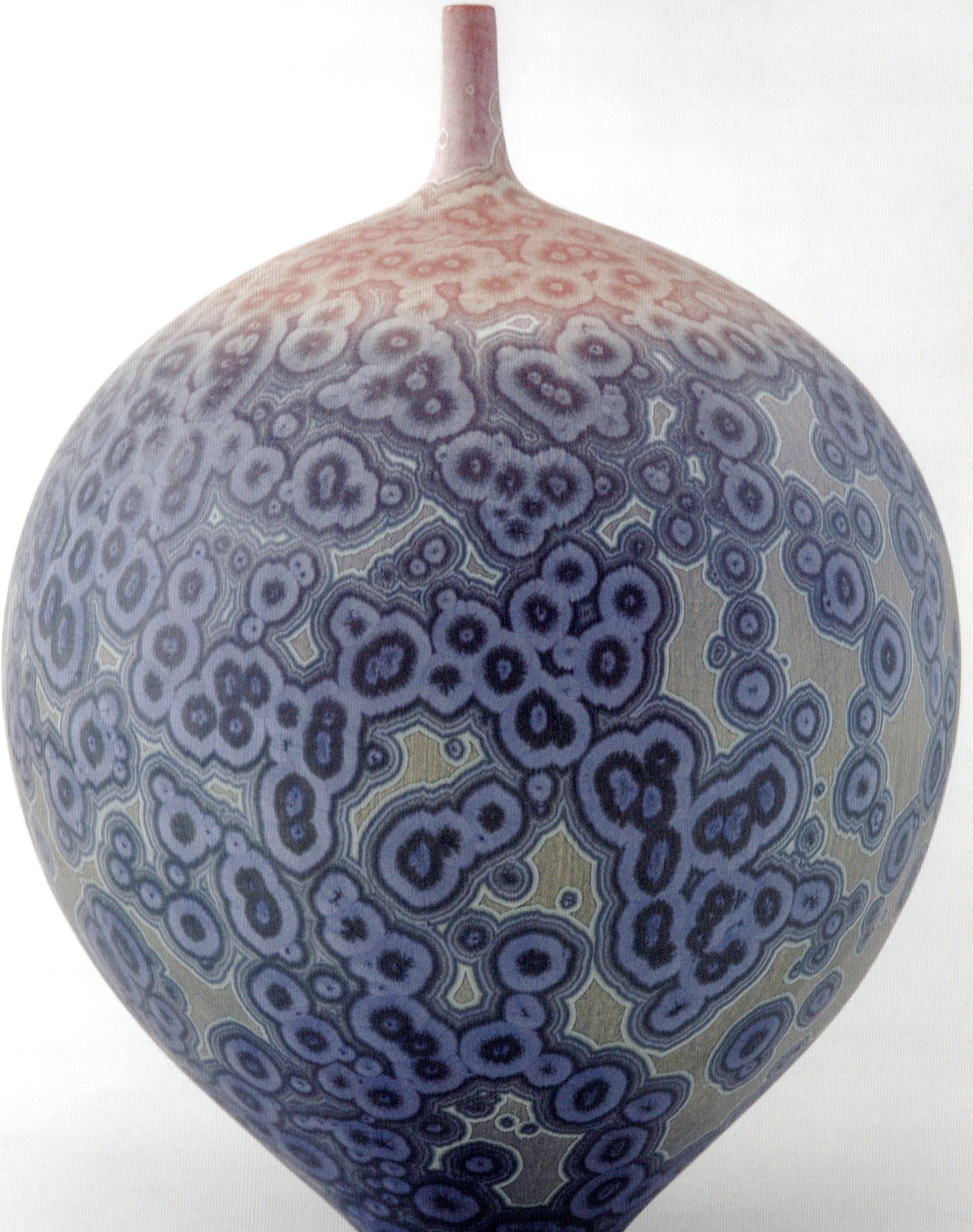

13 Firing crystalline glazes in oxidation

LEFT Matt crystalline © Ted Secombe.

When firing a kiln, leave nothing to chance or the kiln gods. It is essential to have a complete understanding of the procedure, and to keep detailed notes, particularly when working with crystalline glazes.

Creating graph profiles from time studies is tedious, but these profiles provide a fast visual comparison and an understanding of the kiln's firing behaviour. If you aren't using a digital temperature controller, time studies better allow you to repeat a firing programme.

The objective of a kiln firing is to create conditions that promote the growth of crystals. The sequence is:

- Get to maximum temperature.
- Rapidly cool to the crystal-growing range.
- Extended soak, or a slow drop and rise in temperature within the crystal-growing range.
- Shut the kiln off and cool naturally.

Cone 6 crystalline glazes

Bill Schran (USA) found that the kilns at the school where he taught ceramics were not firing to cone 10 quickly enough to grow crystals. He started experimenting by adding increments of low-temperature fluxes and small amounts of lithium carbonate or boron compounds to cone 10 glazes to assist with melting. He was able to get the maximum firing temperature down to cone 6 temperatures.

Bill's firing schedule for cone 6 glazes

RAMPS	TEMP °C	TEMP °F	HOLD
260°C (500°F)/hr	1080	1975	10 min
65°C (150°F)/hr	1196	2185	10 min
260°C (500°F)/hr	1093	2000	10 min
260°C (500°F)/hr	1049	1920	1–4 hr
149°C (300°F)/hr	1010	1850	1–2 hr
38°C (100°F)/hr	816	1500	Off

This firing schedule has cone 6 nearly flat and cone 7 bending.

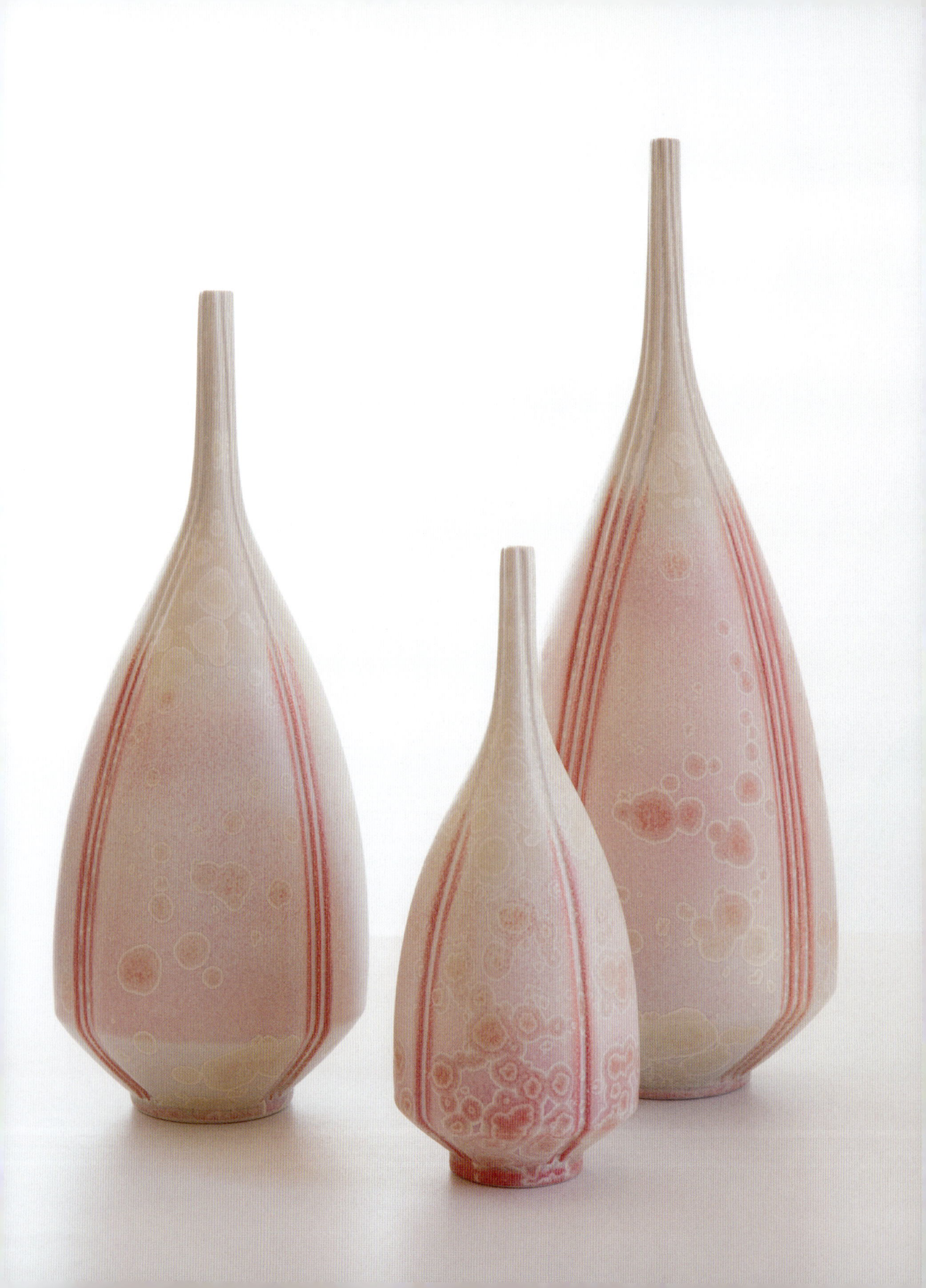

Matt crystalline glazes

LEFT Matt crystalline © Ted Secombe.

BELOW RIGHT Matt crystalline by Hein Severijns.

Some potters prefer many smaller crystals with a subtle blend of colours, the suggestion of depth, and a soft, satiny feel over the flashy, larger crystals. There are several ways to achieve this.

Hein Severijns (Netherlands) felt that large single crystals seduce the observer, and overshadow the complete pot. The softer appearance and silky touch of the matts are more delicate and create better harmony between form and glaze. Hein often polished the surface of his pots with fine carborundum paper to dull the shine and enhance this effect, resulting in a distinct crystal pattern scattered over a silky background.

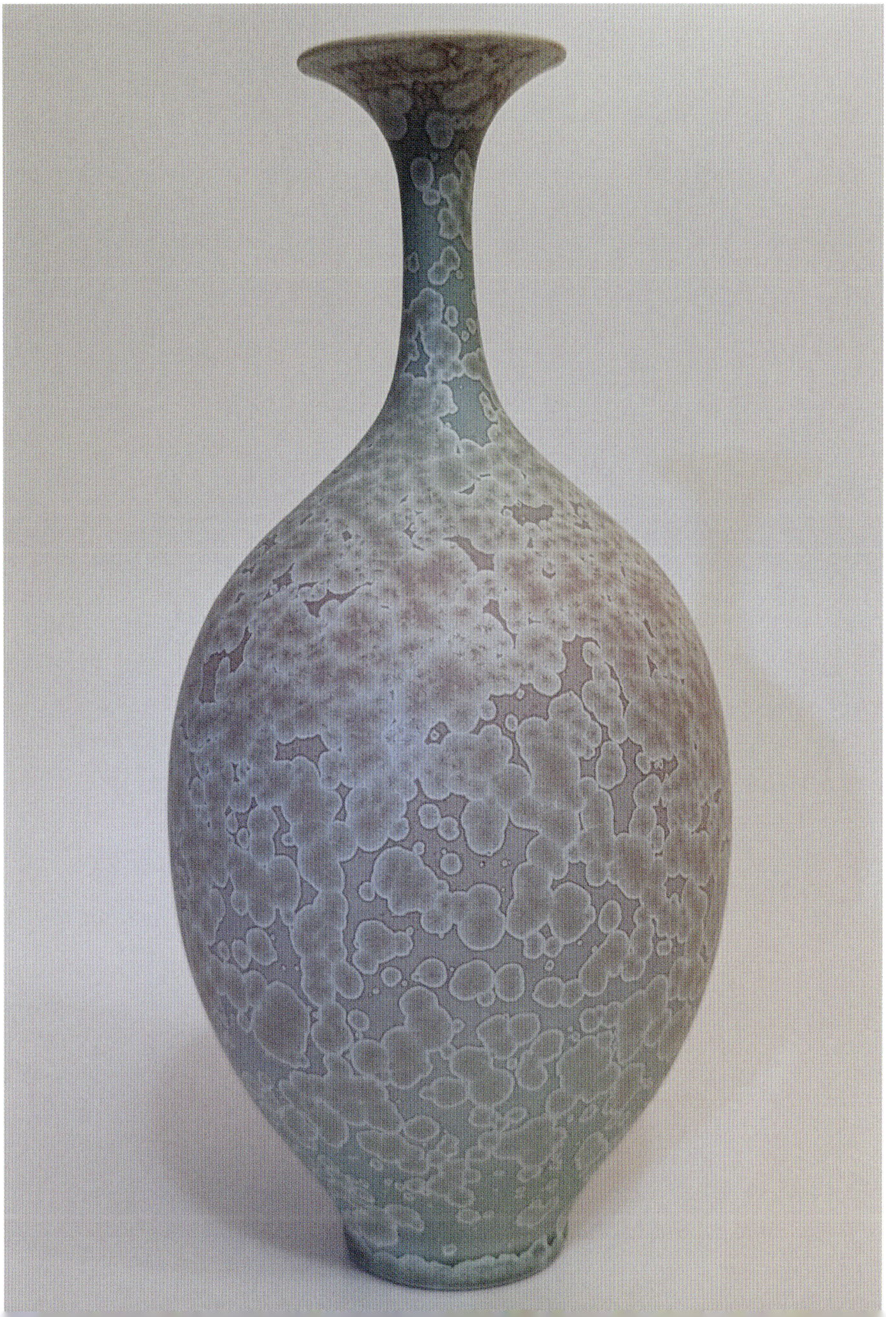

Matt crystalline firing schedule

RAMPS	TEMP °C	TEMP °F	HOLD
Ramp up	1280	2336	0
Ramp down	1130	2066	3 hr
Ramp down	1100	2012	3 hr
Ramp up	1170	2138	2 hr
Ramp down	1140	2084	1 hr
Ramp down	1120	2048	1 hr

BELOW LEFT Matt crystalline © John Tilton.

RIGHT © William Schran.

John Tilton (USA) is also renowned for his crystalline matts. John's firing schedule is given in Chapter 20 (p.182).

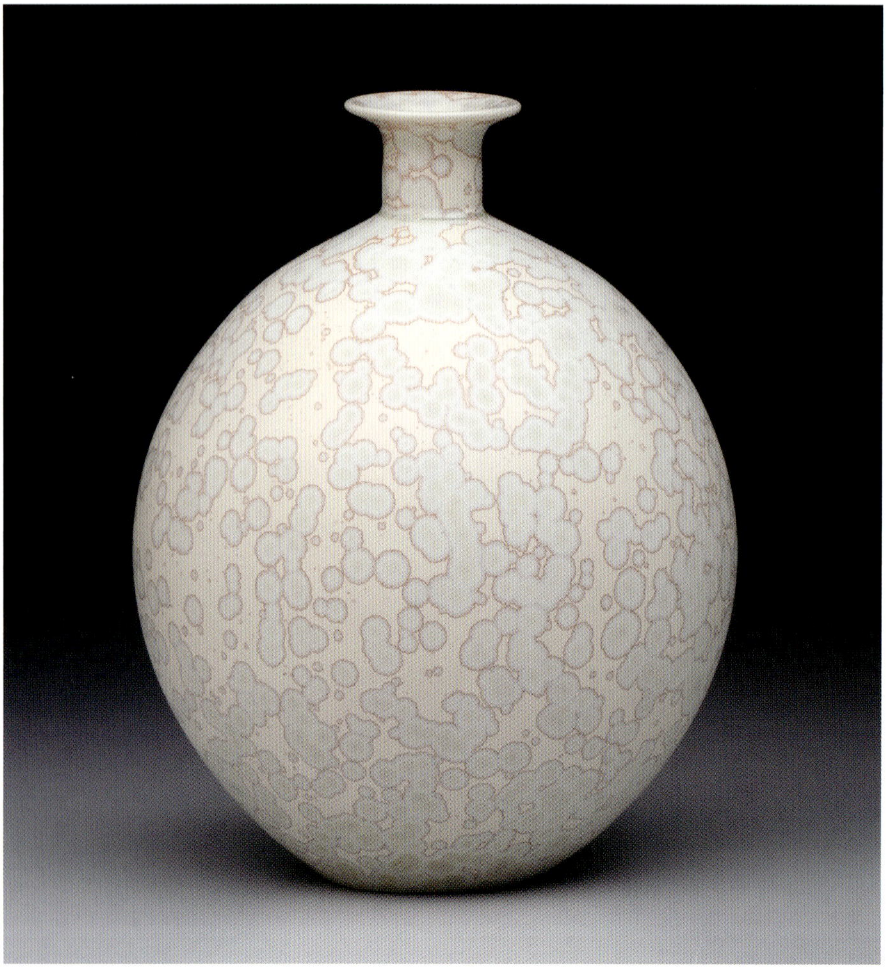

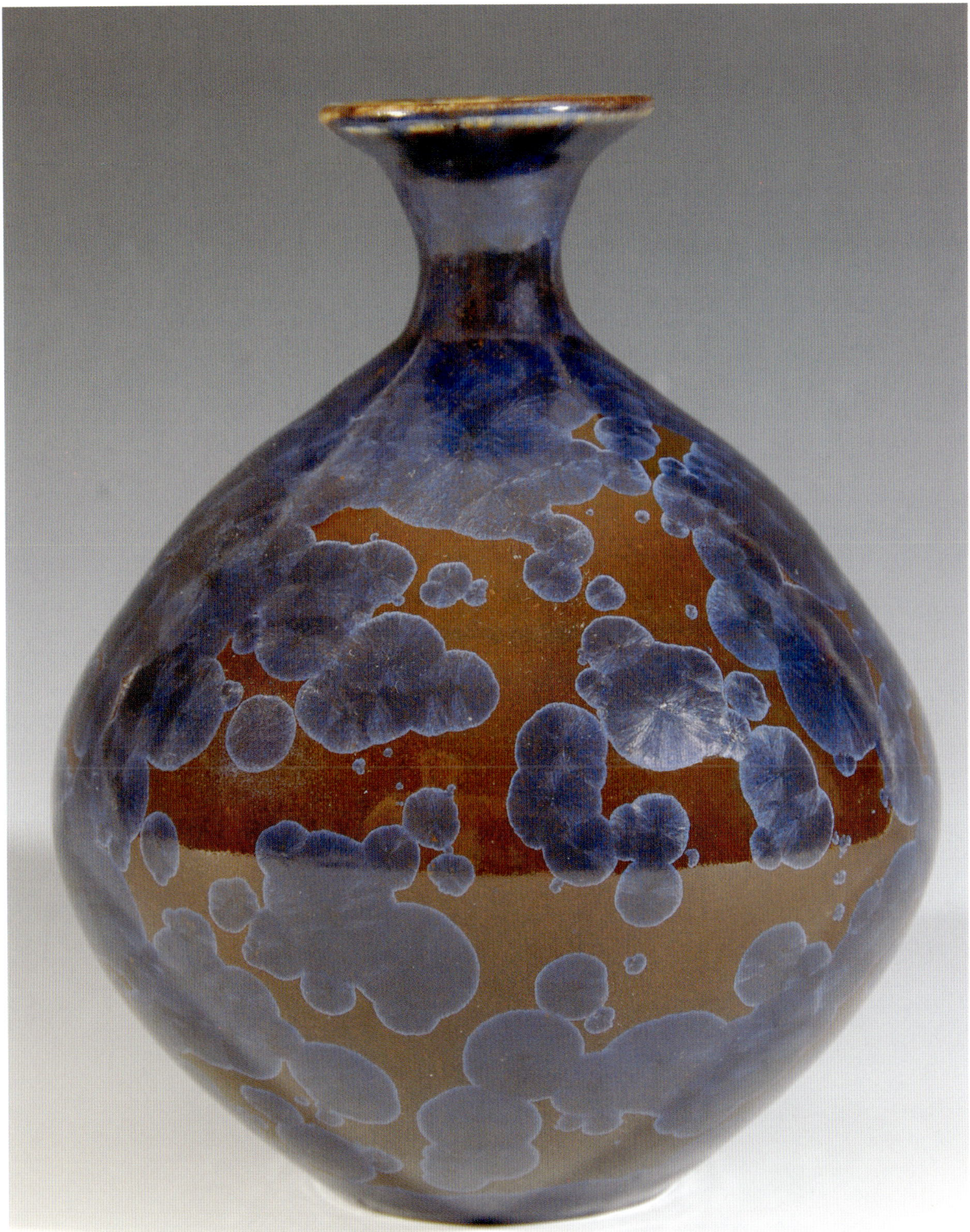

Macrocrystalline glazes, cone 9–11

It is difficult to write a programme for firing crystalline glazes because every kiln fires differently and eccentricities in kilns are more exaggerated at high temperatures. Also, each potter chooses a firing best suited to their specific needs.

The following is a procedure I used when I first began working with crystalline glazes, and only had variable heat switches to control temperatures and a pyrometer to indicate air temperature. The small scale on the pyrometer was difficult to read, and the temperature had to jump significantly before I even noticed. However, this method is a helpful guide. Cones indicated when maximum temperature was reached.

The kiln was turned on in the early evening to low, with spyholes open so moisture and glaze fumes could be vented off without harming the elements. At around 315°C (600°F), both the physical water and most of the chemical water were driven off, so I closed the spyholes and turned the heat up to medium at bedtime.

Early next morning, I turned the temperature to high. By midmorning, cone 10 was bending and maximum temperature was achieved. At this point the kiln was turned off, and I allowed the temperature to drop to the crystal-growing phase.

Once the crystal-growing temperature was reached, as indicated by the pyrometer, the kiln was turned back on, and I maintained the temperature, or allowed it to drop and rise within the crystal-growing range, for a specified time. I never left the kiln during this time, constantly raising the temperature or lowering it to stay within the range. After about four hours, the kiln was shut off and cooled naturally.

Using only a thermocouple and a pyrometer gave a variety of results, from very few small crystals to sometimes large and spectacular ones, but most often the unpredictable results were disappointing. I needed a better temperature indicator.

My next investment was a digital voltmeter attached to the thermocouple wires. Although more accurate, it was still necessary to stay by the kiln during the entire time of crystal growth to monitor the control dials and make changes.

Finally, automatic temperature programmers came on the market. These instruments were a godsend. Not only did they free up valuable time, but certain firing patterns could be created and repeated. It no longer mattered when the kiln was turned on, as there was no need to monitor it, although I liked to be nearby during the later stages of the firing. After the kiln was loaded, I entered a firing programme into the controller and firing log. A firing log should include the following:

- Firing number
- Date
- Glazes used
- Starting time
- Starting temperature (room temperature)
- Rate of temperature climb
- Peak temperature firing to
- Cone achieved
- Rate of temperature drop

- Holding time
- Holding temperatures (ramps)
- Firing variations
- Results

When doing a cone 10 firing with the rate of temperature climb at 150°C (270°F) per hour, the temperature indicated on the programmer, or air temperature, would read 1305°C (2381°F). If the rate of climb were faster, it would be necessary to achieve a higher air temperature to reach the designated cone. This is based on heat work done or the amount of melt, which is cooler than actual air temperature. A slower rate of climb requires a lower air temperature to achieve cone 10. At a rate of climb of 60°C (108°F) an hour, the air temperature indicator would be 1285°C (2345°F). This is necessary to know when indicating the peak temperature of the firing.

The crystal-growing range usually lies somewhere between 1187°C and 1010°C (2170°F–1850°F). Crystal-growing range temperatures are determined through experimentation. The temperatures for each glaze can vary depending on the colourants, as some act as fluxes, but the growing range is usually within 222°C (400°F). The rate of temperature drop to the crystal-growing range depends on your programmer's capabilities. A time study is needed to determine the rate of fall.

Ideally, you want the temperature to drop quickly within the parameters of the programme, and then either remain at a set temperature, or vacillate within the crystal-growing range.

A typical kiln firing would have a slow ramp of 56–112°C/hr (100–200°F/hr) to dry the ware. The second segment can ramp at a faster rate of 112–224°C/hr (200–400°F). The third rate slows again for quartz inversion and carbon burnout between 538°C and 621°C (1000–1150°F). The fourth segment speeds up until 110°C (200°F) below the maturity temperature. The last segment up to maturity may go at 60°C/hr (108°F). Most kilns can achieve this rate, and it matches the rate of rise listed in cone tables. It also allows the glaze to fully melt. These first five segments are easily programmed using the cone fire method of the controllers.

The remaining segments consist of a rapid temperature drop to around 1180°C (2156°F). Once the crystal-growing range is reached, the kiln is held or programmed to rise and fall within the crystal-growing range in order to grow halos. Many complicated firing schedules with unique results can easily be achieved with modern controllers. If you are seriously interested in doing many firings, you should consider investing in a digital interface processor. The technology allows freedom to pursue other activities, instead of monitoring the kiln.

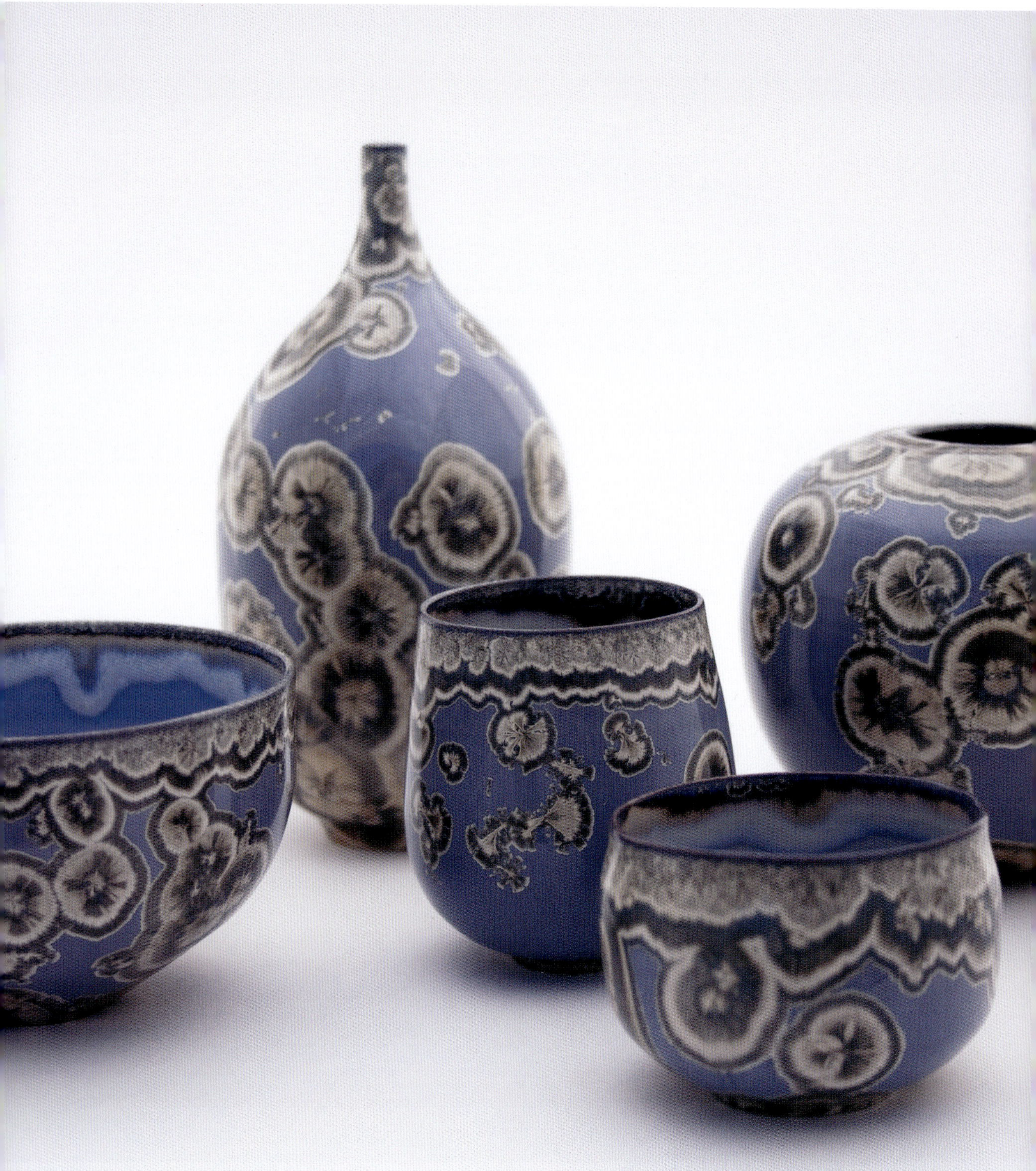

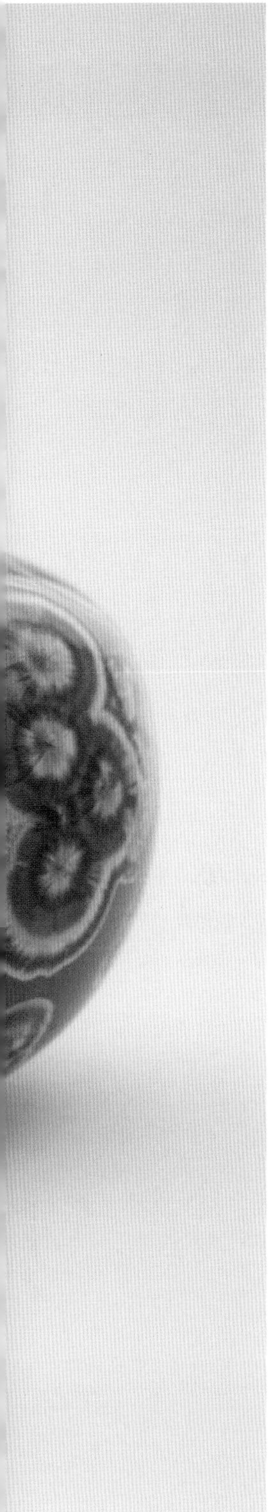

Halos

Once the crystals have started growing, raising and lowering the temperature throughout the soak time creates growth rings, or halos. With each rise and fall, a concentric ring forms around the outer edge of the crystal, like ripples when a pebble is tossed into a pool. Placement of rings can also be controlled, depending on when in the crystal-growing cycle the rises and falls occur.

Controlling the crystal shape

Different shapes can be achieved at different soak temperatures. Rods or needles are achieved by soaking at the hotter end of the crystal-growing band; a few degrees cooler produces Maltese cross shapes; cooler still, and the crystals will resemble fireworks exploding. However, to achieve a particular crystal shape, the temperature must remain constant within the growing range.

For those who do not have a programmer for exact temperature control, keeping the temperature within 55°C (100°F) of the target temperature usually produces nice crystals. A soak time of two hours or less produces small to medium-sized crystals; three to four hours produces larger ones; five to six hours produces even larger crystals, but often they butt up against one another. When crowded together, their growth is restricted, becoming a mass of crystals with a less spectacular appearance.

This temperature outline is not specific. It is necessary to experiment with each individual glaze to find its most favourable growth patterns. Once the appropriate fusion temperature is determined, variations in the growth phase can be performed. This will provide an understanding of optimal conditions for crystal formation.

If you are converting temperatures from Celsius to Fahrenheit or vice versa, using a calculator or your phone with conversion capabilities will give you the equivalent temperatures, or you can convert using the following formulae:

Celsius to Fahrenheit:
(°C x 1.8) + 32 = °F

Fahrenheit to Celsius:
(°F – 32) ÷ 1.8 = °C

However, if you are converting the rate of temperature rise or fall, or are describing the difference between two temperatures, different formulae must be used.

Celsius to Fahrenheit:
°C x 1.8 = °F

Fahrenheit to Celsius:
°F ÷ 1.8 = °C

This is necessary to know when deciding on rate of temperature climb or drop, or describing the increments between two temperatures, and it is necessary to convert from one scale to another.

LEFT Example of halos © Annie Cap.

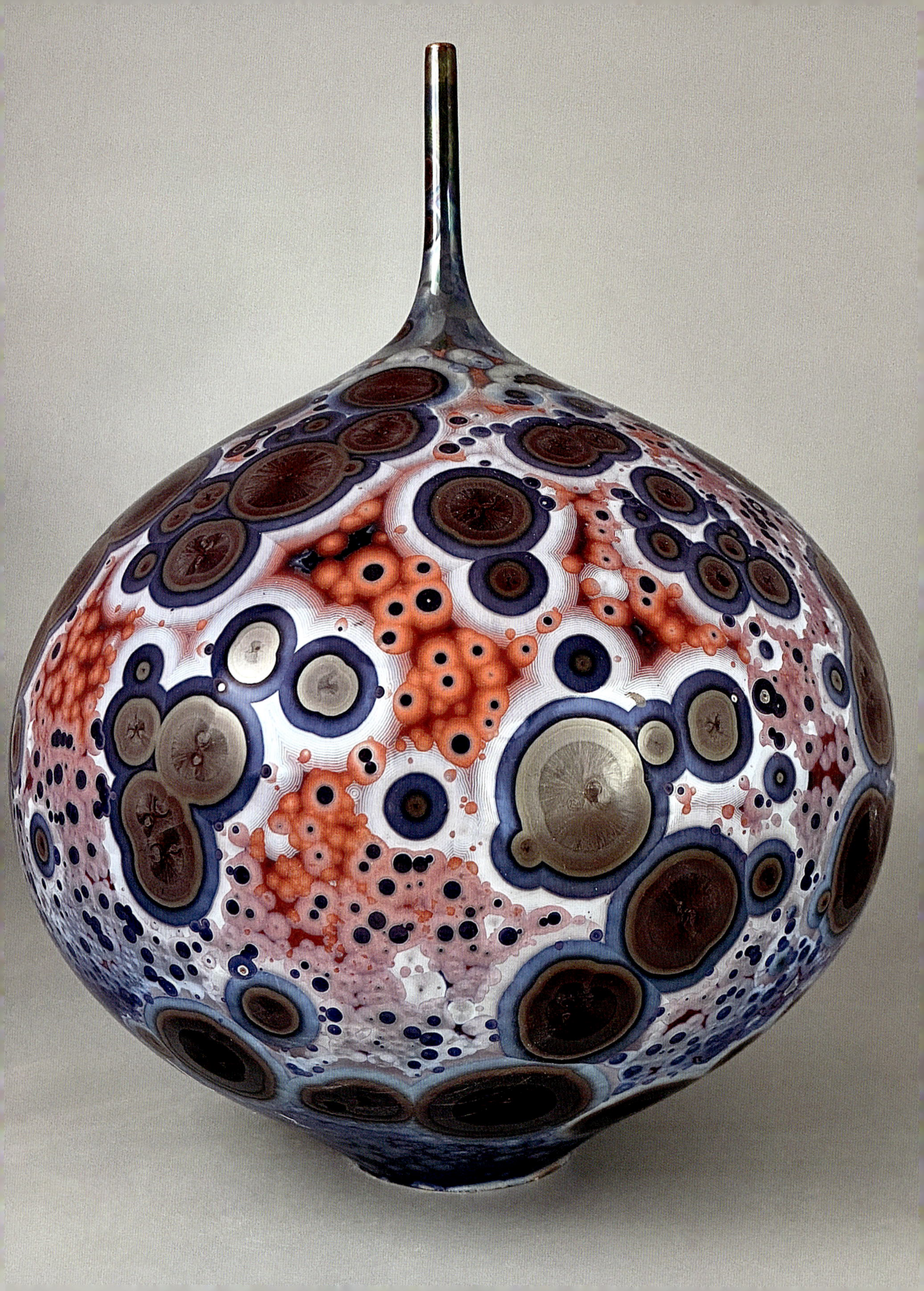

14 Reduction-fired crystalline glazes

LEFT Bottle fired in reduction © Jose MARISCAL Paneque.

Before electric kilns were available, crystalline glazes were fired in fuel-burning kilns. To shield certain glazes from reduction, pottery was placed in saggers (a boxlike clay container used to protect the ware). Since carbon adheres to and is absorbed by the clay and glazes, it alters fluxing action and dramatically changes the colours of both. This is called a 'reduction' atmosphere (reduced oxygen to fuel).

Crystalline glazes contain a large proportion of zinc oxide compared to standard glazes. The reduction of zinc oxide begins at 1025°C (1877°F). Zinc oxide reacts with a reducing agent, carbon, to produce zinc and carbon monoxide. When the temperature reaches 1265°C (2309°F), almost 100 per cent of the zinc is reduced and vaporized.

Even slight reduction is sufficient to extract the oxygen, which results in the loss of the zinc. Therefore, any reduction at these or higher temperatures will leave a boiling metal which volatilizes and is lost. It was widely advised that glazes with zinc oxide be fired only in fully oxidised or neutral firings, which is why the long-held belief existed that crystals could not be grown in a reduction atmosphere.

Many potters are now realizing that firing crystalline glazes in reduction adds a whole new dynamic to their glazes. Most metallic oxides react differently in a reduction atmosphere. For example, copper in oxidation is green, while in reduction it turns a brilliant red. Glazes containing iron are enhanced both in colour and fluxing action by reduction. A clay body that has iron in its composition, when fired in reduction loses its paleness and becomes a toasty colour. Even porcelain may go from a creamy colour to a slightly translucent white.

When growing crystals, the kiln must first be fired in oxidation, with reduction taking place during the later part of the crystal-growing phase, or later still, during cooling. Reduction can be performed in both electric and gas-fired kilns.

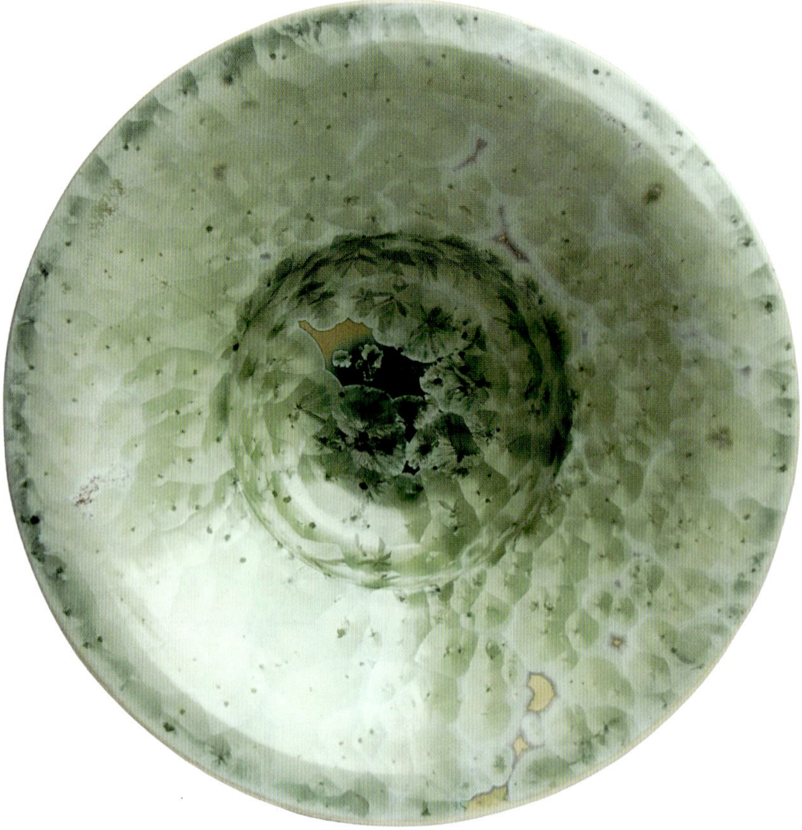

LEFT Plate fired in oxidation.

BELOW LEFT Same glaze fired in reduction.

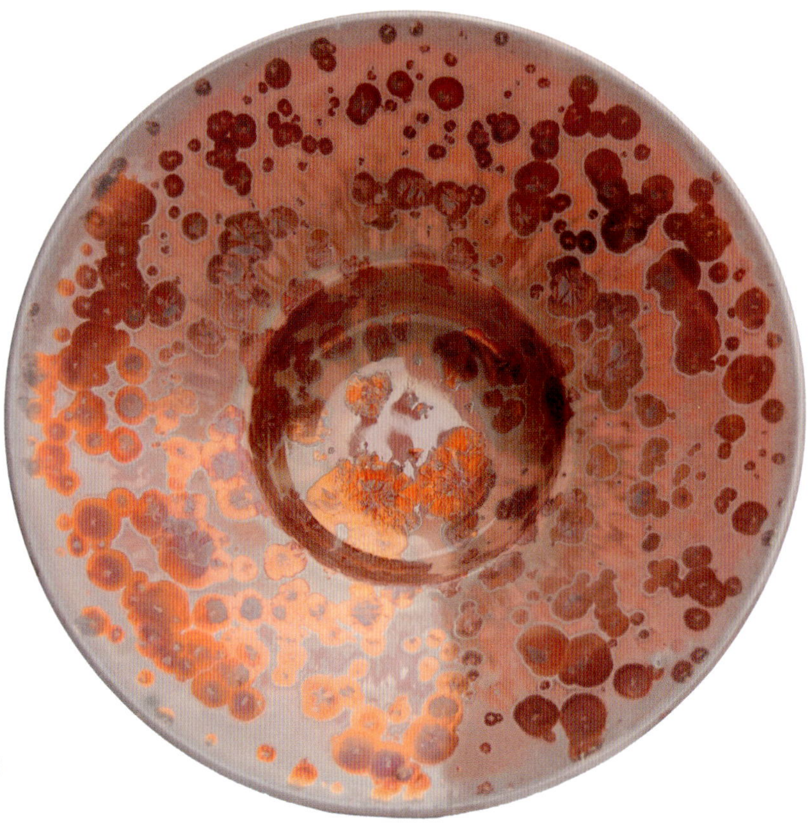

Electric kiln reduction

Before even considering reduction in an electric kiln, make sure you have a carbon monoxide detector and a fire extinguisher nearby. If you have a proper exhaust system and wish to try this procedure, there are several ways it can be done.

The pots are first fired in oxidation to grow the crystals, and then the kiln is shut off. When cooled to 815°C (1500°F), reduction is introduced while the kiln cools further until the temperature drops to 538°C (1000°F), at which point reduction is discontinued.

Alternatively, a separate reduction firing can take place. Previously fired pots are loaded into the kiln and the temperature is taken to 815°C (1500°F). The kiln is shut off and reduction occurs as the temperature drops to 538°C (1000°F).

A combustible material is needed to cause reduction. Sugar cubes or small pine chips inserted into the bottom bung of the kiln will produce carbon. This method is difficult to control, and usually results in spotty reduction and inconsistent outcomes. It also degrades the electric kiln elements prematurely.

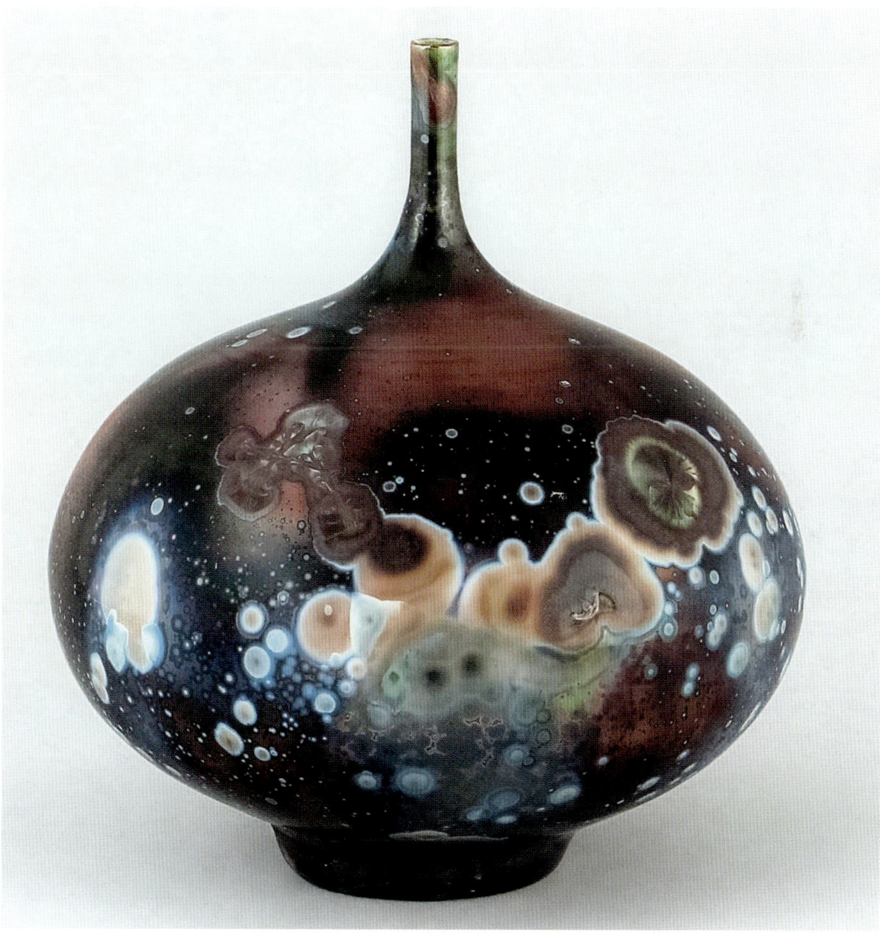

RIGHT Fired in reduction © Isak Isaksson.

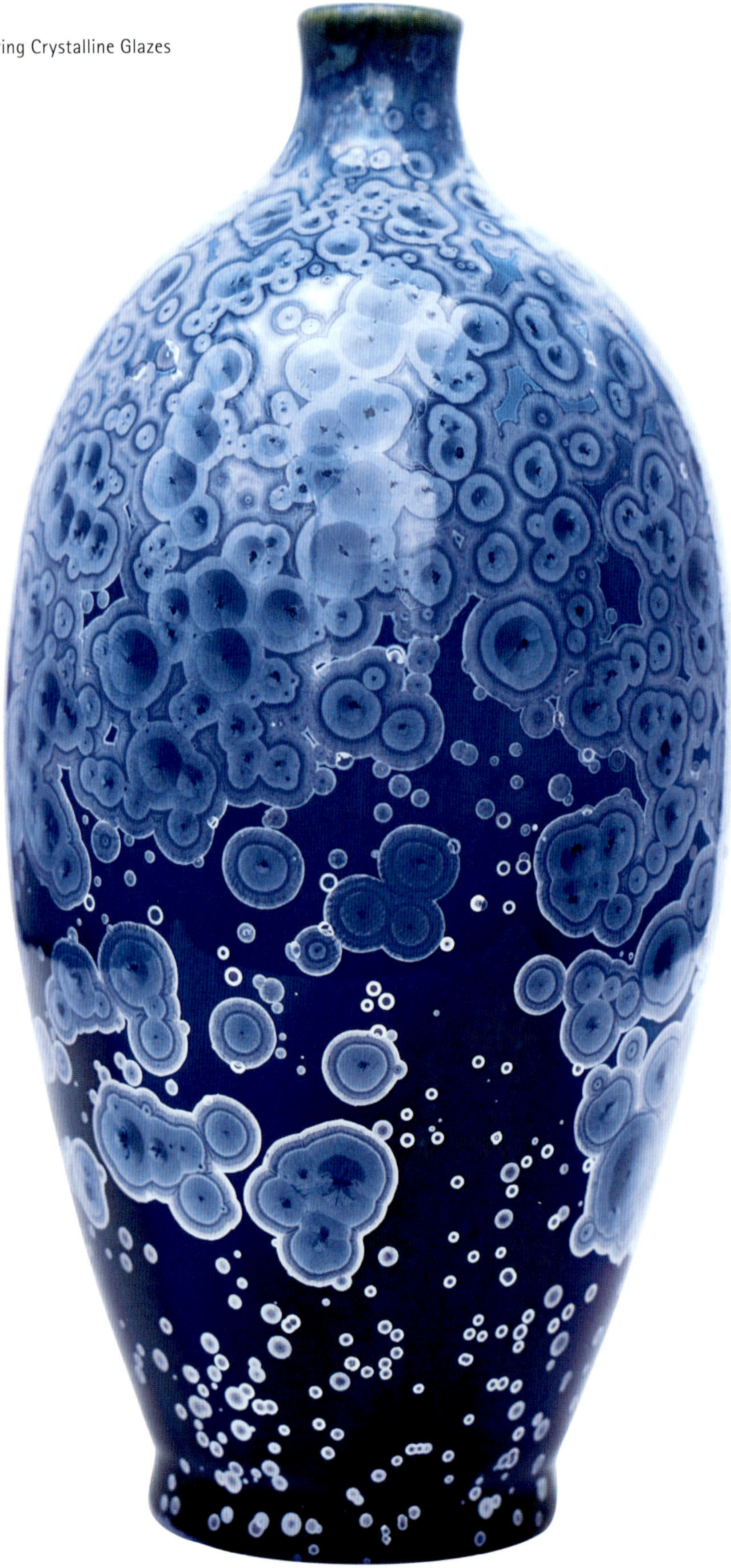

LEFT Bottle, fired in oxidation © Casey Anderson.

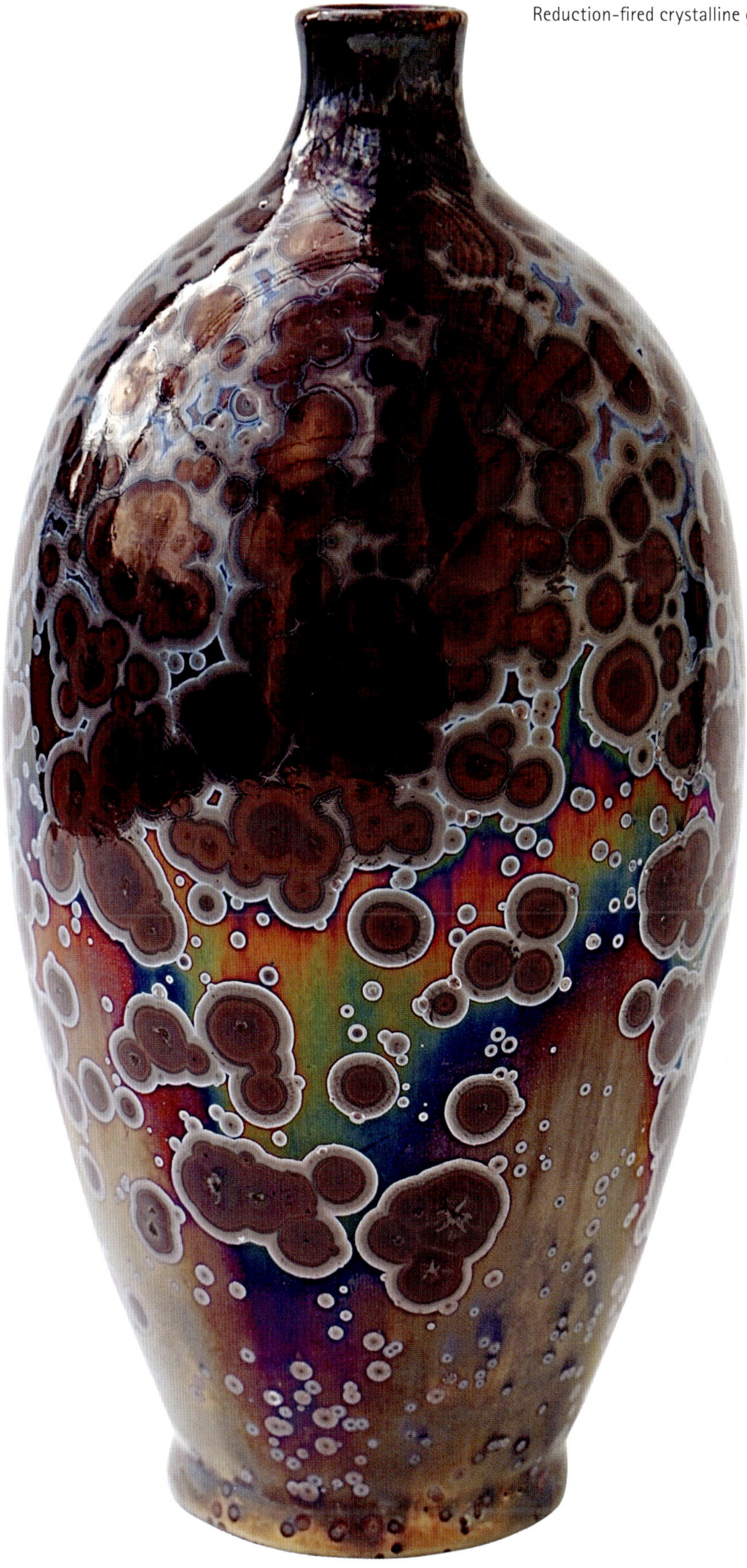

RIGHT Same bottle fired in reduction © Casey Anderson.

A more reliable method is introducing fuel by way of a two-litre IV bag with a surgical tube attachment, available from some pharmacies or a hospital supply store. The bag is filled with cooking oil or denatured alcohol. A flow control that regulates the drip is attached to the tube. The kiln is fired (or allowed to cool down) to around 815°C (1500°F). The IV bag is suspended above the kiln with the end of its tube inserted into a protective ceramic sheath that enters the kiln through the bottom bung. The point of entry into the peep hole should be sealed with clay or fibre blanket to hold the sheath in place and make the area airtight so that complete combustion takes place.

With the kiln off, oil (or alcohol) is introduced at around 815°C (1500°F). The pinch valve is adjusted to allow the fuel to enter the kiln at approximately one drop per second. Because of the lack of oxygen, carbon is created. After the kiln has cooled down to 538°C (1000°F), the IV bag is removed.

RIGHT Covered jar. First fired in oxidation then post-fired in reduction.

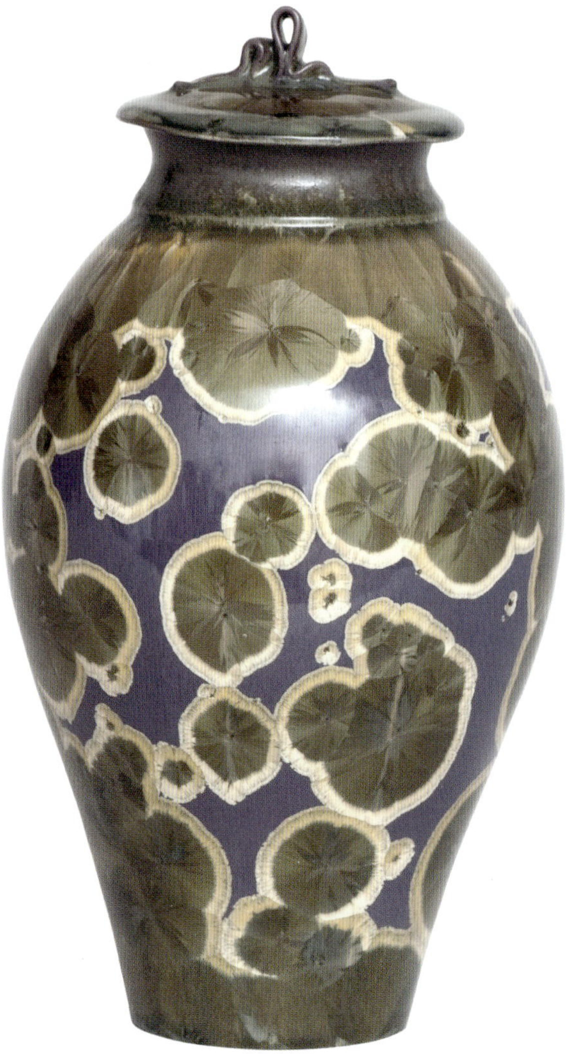

RIGHT First fired in oxidation then post-fired in reduction.

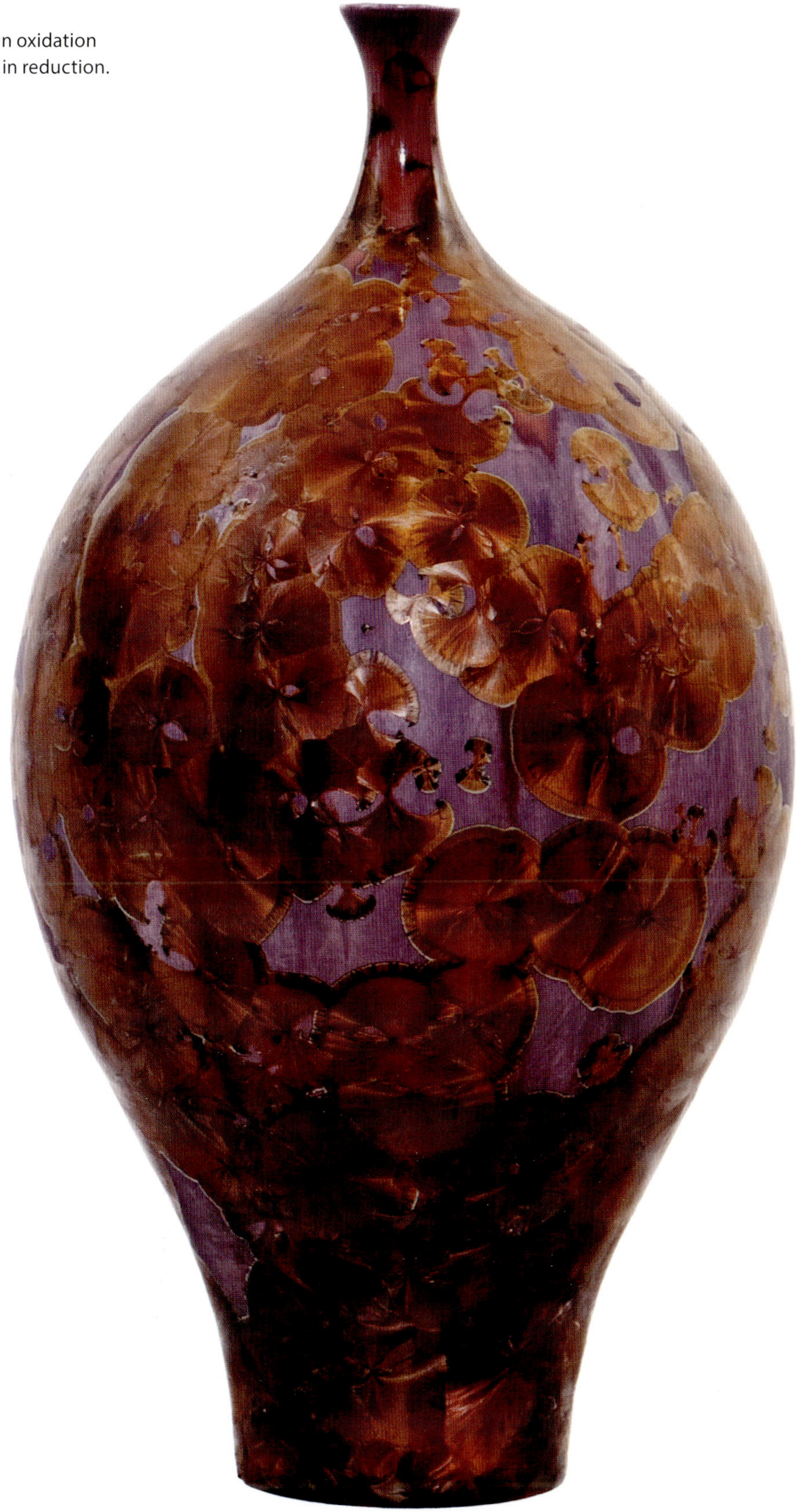

Reduction is determined by peeking into the upper bung. A hazy orange glow or small flame with wispy carbon trails should be visible. If a larger flame escapes, the reduction is too heavy. A clear atmosphere tells you reduction should be increased.

In the past, potters generally thought that extended reduction in electric kilns led to element breakdown, increased repair costs, and carbon build-up on the kiln interior, but improved element quality is challenging this belief. Higher temperatures do increase stress on the elements, but using the electric kiln for bisque firings re-oxidises them and burns away the soot.

For a different effect, reduction can be introduced at higher temperatures during the later part of the crystal-growing phase, starting around 1093°C (2000°F) and reducing

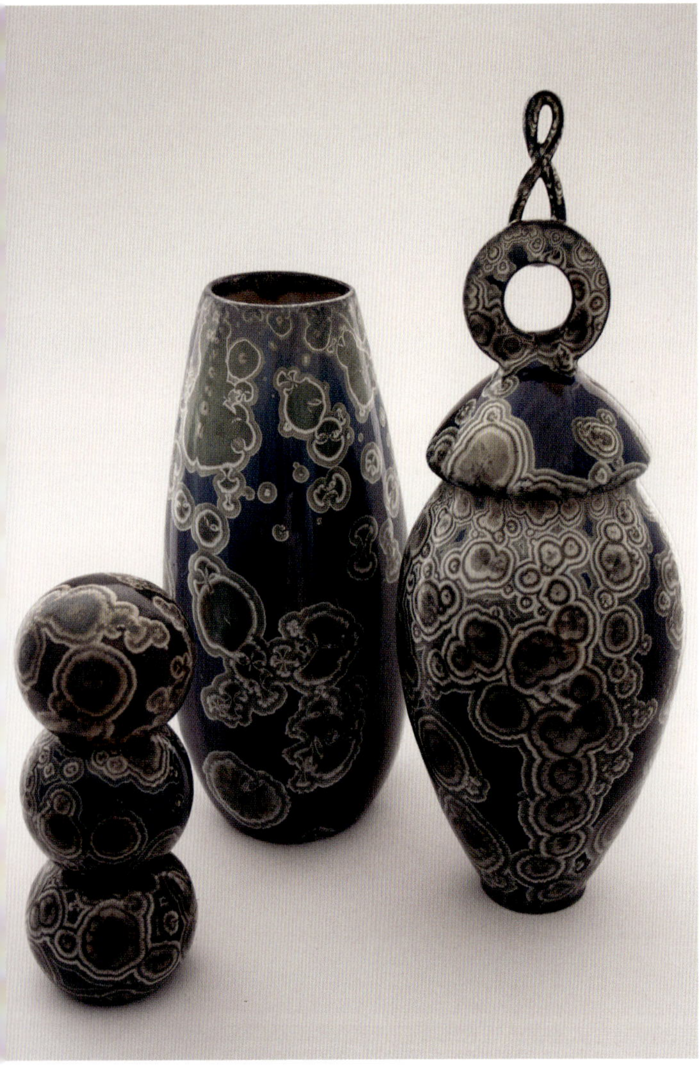

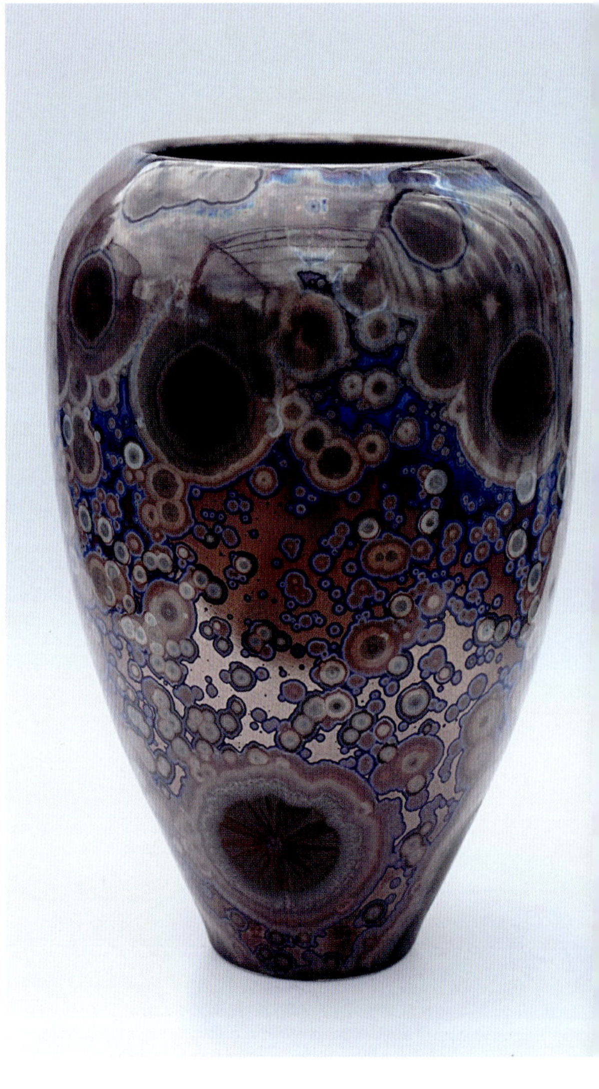

to 1010°C (1850°F). Alternatively, you can hold the temperature at 1093°C (2000°F) and reduce for about 30 minutes during the last half hour of the crystal-growing phase.

Annie Cap (England) introduces reduction into her electric kiln between 1111°C and 1046°C (2032–1915°F) during the later stages of crystal growth. Annie uses an IV bag and pure methanol as her fuel. She believes a more vibrant colour palette is achieved by reducing in the higher temperature range.

Some potters have electric kilns with gas reduction capabilities; others post-fire the pots in a gas kiln. Even doing a post-reduction firing in an airtight raku kiln is a possibility. Regardless of the different ways to introduce reduction, know the firing capabilities of your kiln.

FAR LEFT Post reduced at the higher temperatures © Annie Cap.

LEFT Reduction fired pot © Casey Anderson.

RIGHT Reduction fired © Jose MARISCAL Paneque.

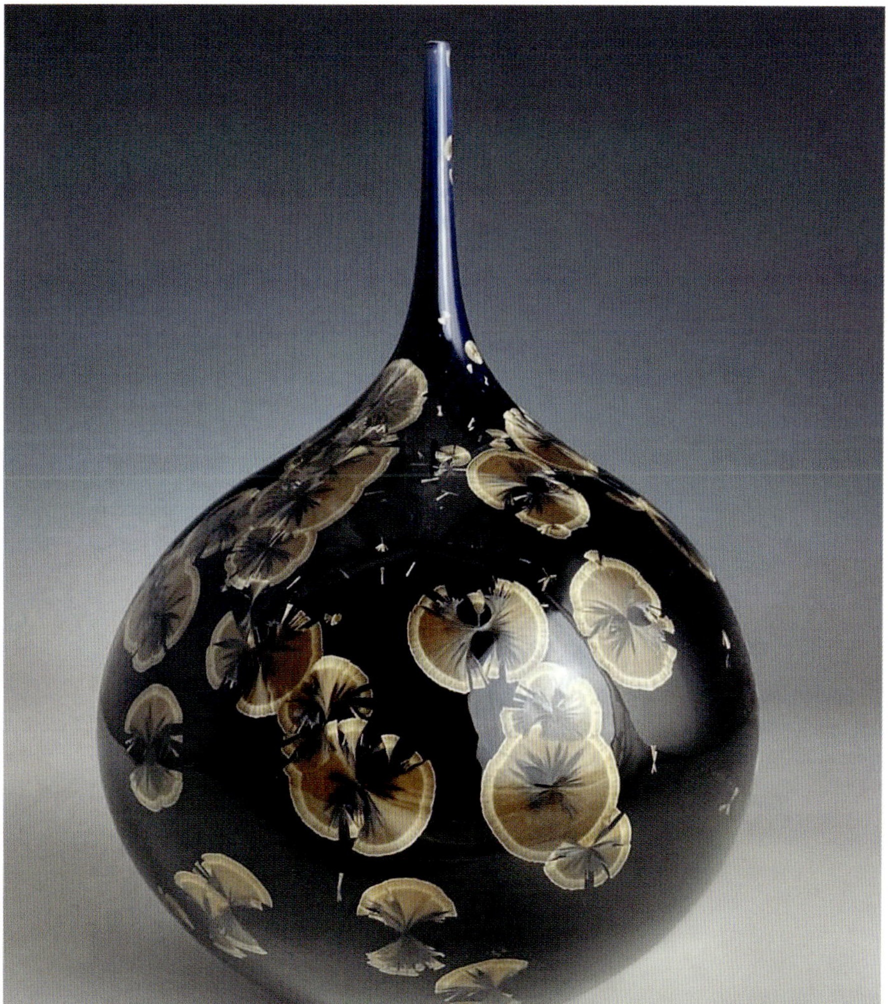

Gas-burning reduction

Safety must always be a factor when firing kilns, especially where fuel is involved. Gas kilns, because of their toxic emissions, should be housed out of the studio, preferably in a separate structure with excellent ventilation. Reduction should not be attempted unless there is a hood to extract fumes, or the firing is conducted outdoors.

Modern gas kilns are designed to control the atmosphere. For firing crystals, the kiln is first fired in neutral or oxidation. The damper must be adjusted substantially to support sufficient air supply for on-ratio combustion (proper ratio of gas and air supply for complete combustion). Using an oxygen sensor, (an instrument that monitors the oxygen present in the atmosphere), the potter can ensure that no reduction takes place during the initial crystal growing. Reduction is introduced after crystal growth or during the later stages. Similarly to electric kiln reduction firings, different results occur contingent on the temperature and timing.

Some gas kilns allow you to pre-programme the complete firing in both oxidation and reduction.

One glaze can give three dissimilar results depending on the firing schedules. For instance, a glaze known as Gold Stuff (see Fig. 1), fired in oxidation, gives gold-coloured crystals on a light tan background. Reducing at higher temperatures results in a blue background with gold-coloured crystals ringed in white (see Fig. 2). The same glaze reduced at lower temperatures produces a deep purple background with olive green to gold crystals ringed in gold (see Fig. 3). It is the firing that gives the diversity, with distinctive results depending on both timing of reduction and the chosen temperatures.

Gold Stuff Glaze: Cone 10

INGREDIENT	PARTS
Silica	28.50
Fusion Frit F-644	45.10
Zinc oxide	25.30
Bentonite	0.70
Molochite	0.40
Total:	100.00
Add:	
Rutile	5%
Ilmenite	3%

Reduction of crystalline glazes offers the potter even more variety to an already vibrant glaze process. A whole palette of amazing colours can be achieved simply by altering the firing conditions.

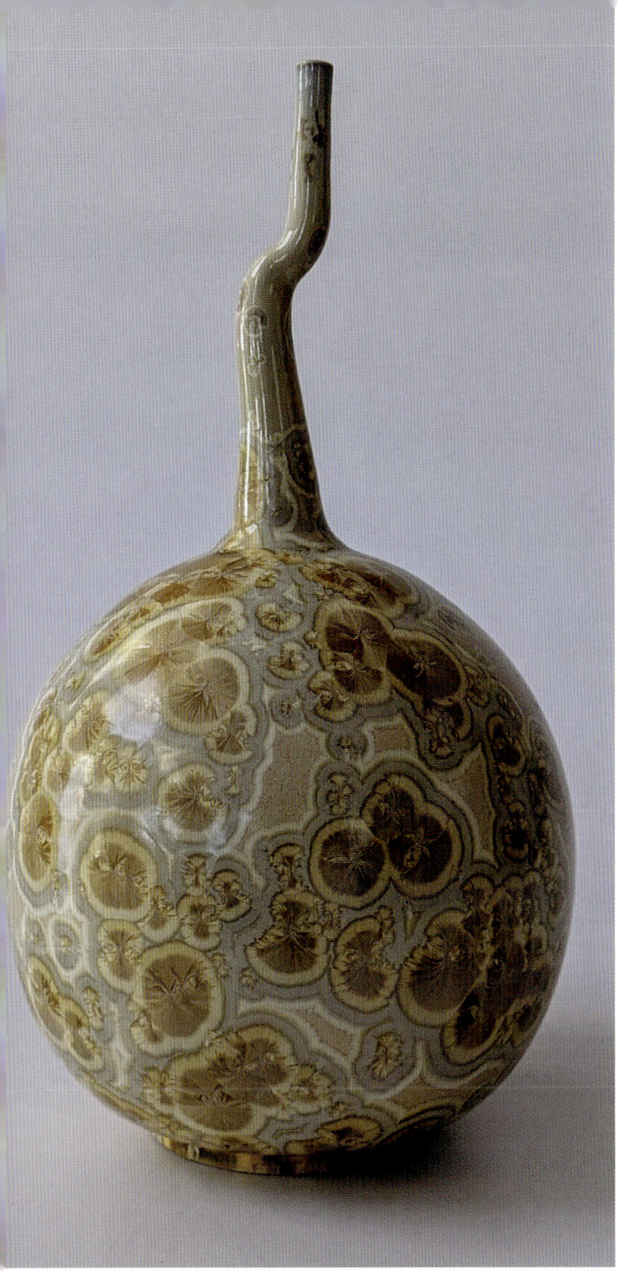

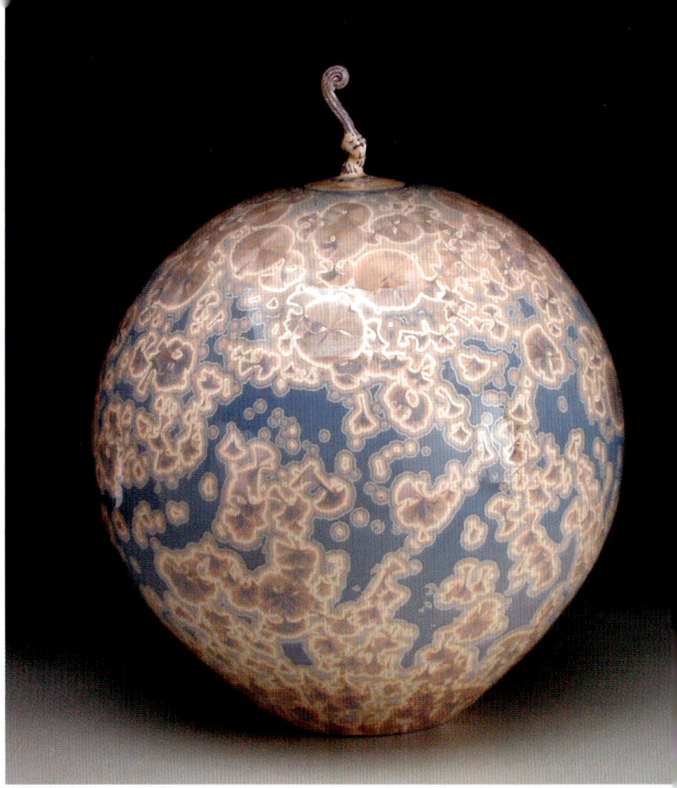

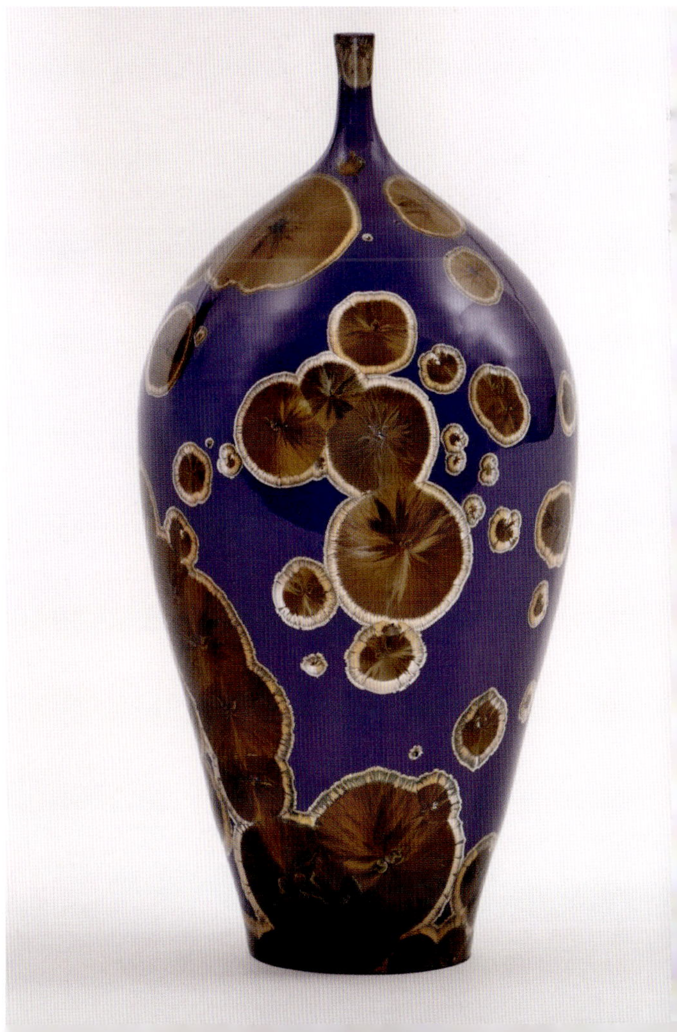

ABOVE Fig. 1: Gold Stuff glaze, oxidation-fired.

ABOVE RIGHT Fig. 2: Gold Stuff glaze first fired in oxidation then reduced at higher temperatures © John Tilton.

RIGHT Fig. 3: Gold Stuff glaze first fired in oxidation then reduced at lower temperatures.

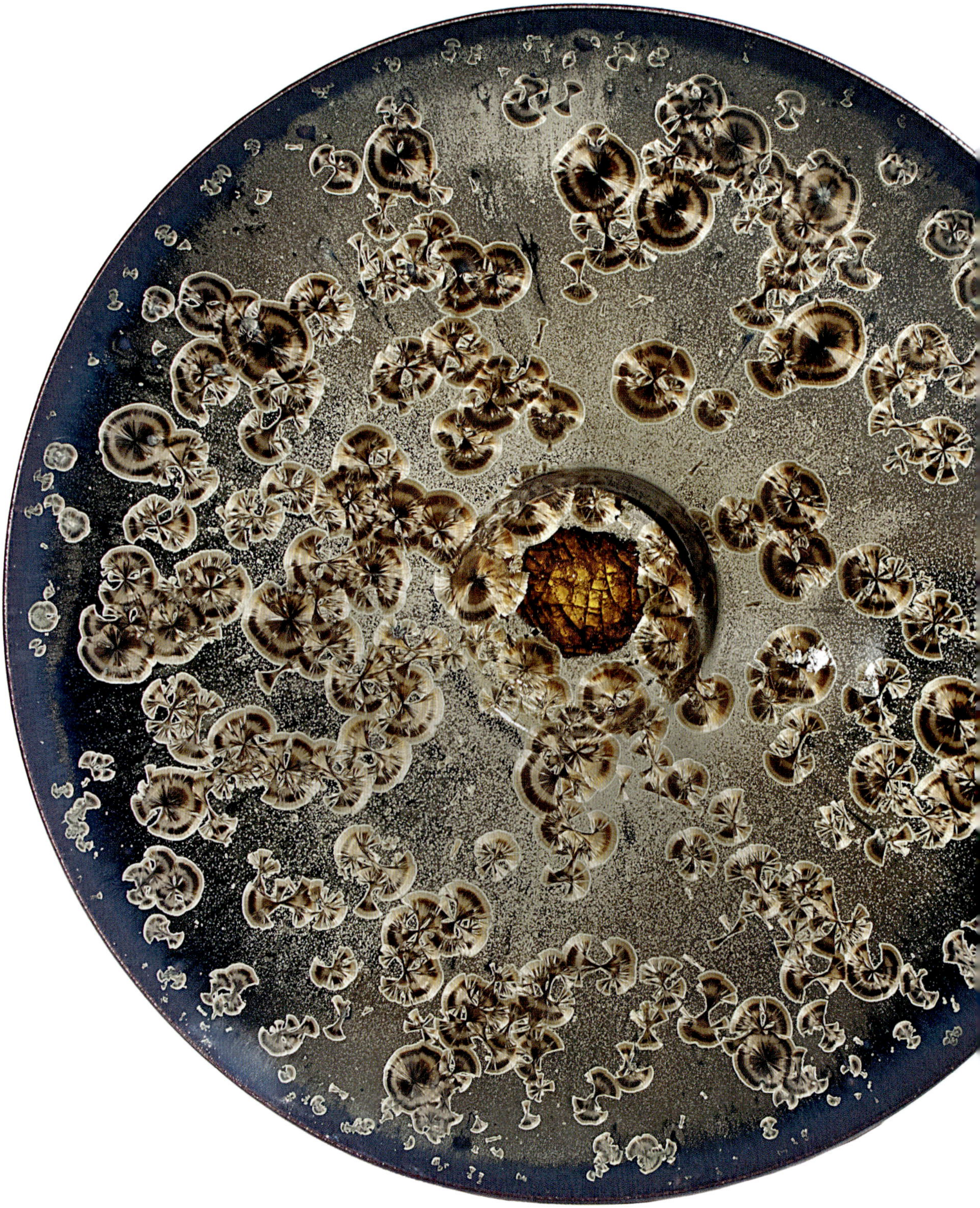

15

After the firing

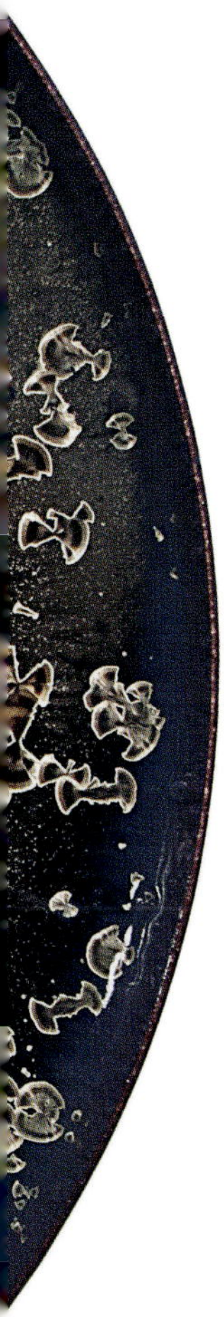

Opening the kiln is always a surprise. We are joyful at our successes; we learn from our mistakes. This is where the notebook comes in handy as we compare results and draw conclusions.

After the kiln is shut off, it should cool naturally. This facilitates a longer life for the kiln and elements. The temperature should drop to 210°C (410°F) before opening the door or spyholes. Wait until you can lift the pots with your bare hands before unloading.

Pot, pedestal, and collection dish are usually stuck together, so when picking up, lift by the collection dish, not the pot. I have picked up a pot with the collection dish still attached, only to have it separate mid-air and fall back into the kiln, breaking other pieces.

Removing pots from pedestals

Sometimes just lightly tapping the catch basin on a tabletop breaks the pot away from the pedestal. If this doesn't work, here are a few suggestions:

- As the pot turns on a banding wheel, direct a small handheld propane torch with a blue flame at the point where the foot rim is attached to the pedestal. The heat usually causes the pot to break free.
- Run a diamond-tipped glass cutter or a tungsten tile cutter around the seamline where the bottom of the pot joins the pedestal. A diamond glass cutter is more expensive than a regular glass cutter, but it never dulls and is capable of cutting deep into the glaze, breaking the bond.
- If the pedestal is still attached, don safety goggles and, using a small hammer, tap the seamline gently with a 13 mm (1/2 in) titanium-tipped chisel or drill bit after scoring with the glass cutter. Many light taps around the seam should break it free.
- Be careful about tapping too hard, and don't tap repeatedly in one area. Teardrop forms are notorious for breaking above the foot rim where the clay is thinnest. These methods should separate the pot from the pedestal, although sometimes the glaze has seeped between the joints, and the pot and pedestal remain stuck together. Then, you will need to consider other methods.

LEFT Plate © Bill Boyd.

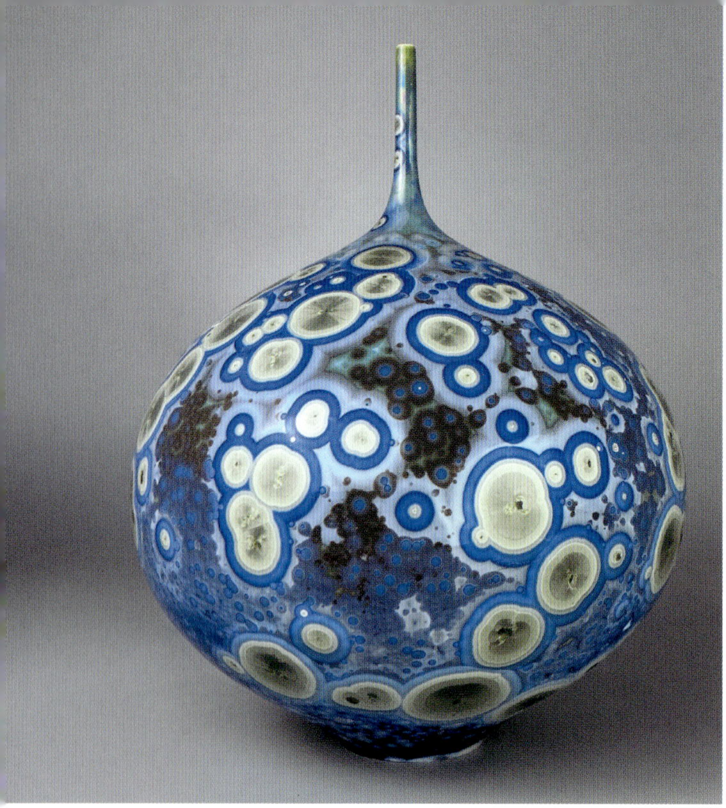

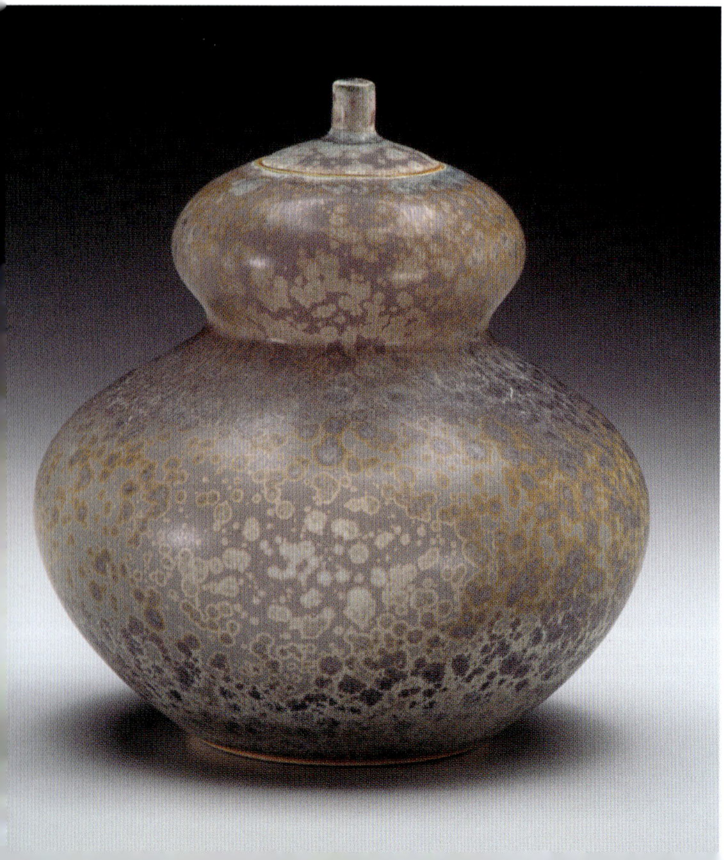

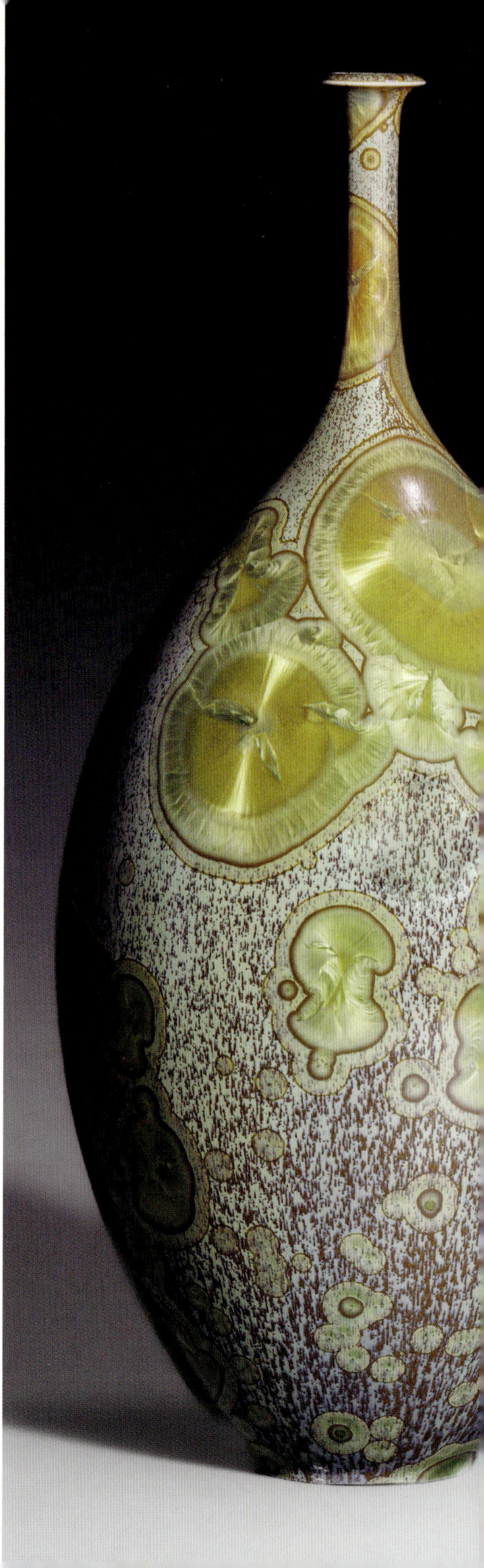

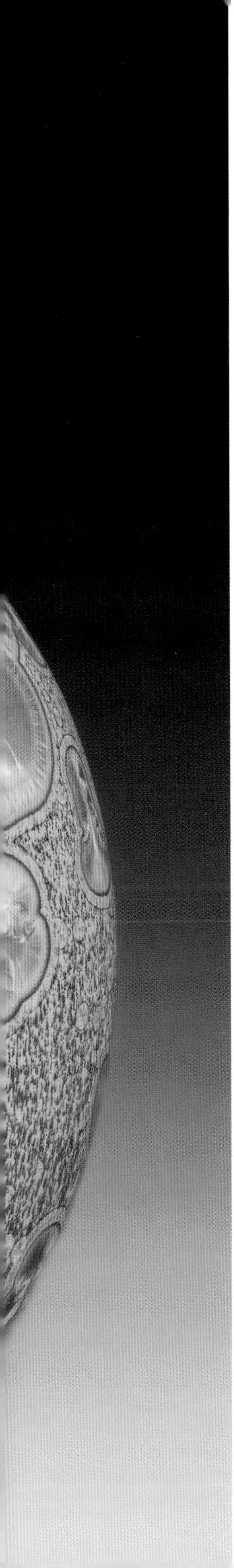

John Tilton uses an electric diamond saw, 25.5 cm (10 in) in diameter and about 3 mm (1/8 in) thick. As the saw spins, he makes a ridge just under where the pot is glued to the pedestal. He keeps cutting until the pot and pedestal separate. Necessary safety precautions must be taken. I don't have a diamond saw, nor do I feel confident handling one, so if I reach this point and still I can't separate pot and pedestal, I toss it out.

After pots are removed from their pedestals, the bottoms can be razor sharp, and can cut your hand or scratch a tabletop. Sometimes, unsightly marks are left on the bottom. A silicon carbide grinding wheel can be used for grinding and smoothing. Wear safety goggles, and hold the bottom of the pot to the spinning wheel, pressing lightly, and with a slight rocking motion tilt the pot around on the outer edge of its foot rim.

If you don't have a silicon carbide grinding wheel, a silicon carbide disc can be glued to a throwing batt and attached to the wheelhead of the potter's wheel. With the wheel turning at highest speed, the pot is pressed gently to the disc.

Don't use an aluminium oxide grinding wheel. It can create too much heat at contact points, sometimes causing the pot to shatter. It may also leave a dark, dirty-looking bottom. Jewellers' grinding stones (14,000 rpm), lapidary grinding wheels, or glassblowers' equipment designed for levelling bases also work well.

A diamond disc is even more effective. Water dripping slowly onto the disc lubricates it as the wheel turns. I don't have a method to drip water continuously, so I placed a bowl of water beside the wheel and would occasionally squeeze a damp sponge onto the disc.

Lap wheels are tools made specifically for the purpose of smoothing the bottoms of glass. They come with a motor, diamond discs of various grits, and a way to stream water onto the disc. John Tilton uses a Crystalite® CrystalMaster™ Pro 12 with a 100-grit diamond disc. He's found this is very effective for taking off the sharp glaze around the outside foot, and for smoothing the bottoms.

John also has an expanded rubber wheel which holds a silicon carbide belt and is used for rounding the edges of the bottoms of pots so that there is a slight shadow underneath. It is used after the lap wheel to further finish the bottoms.

Pots must sit flat on a tabletop. Dirty spots, rough bottoms and ineffective grinding detract from the overall quality of the finished piece.

TOP LEFT Bottle © José MARISCAL Paneque.

FAR LEFT Covered jar © John Tilton.

LEFT Bottle © Richard Bideau.

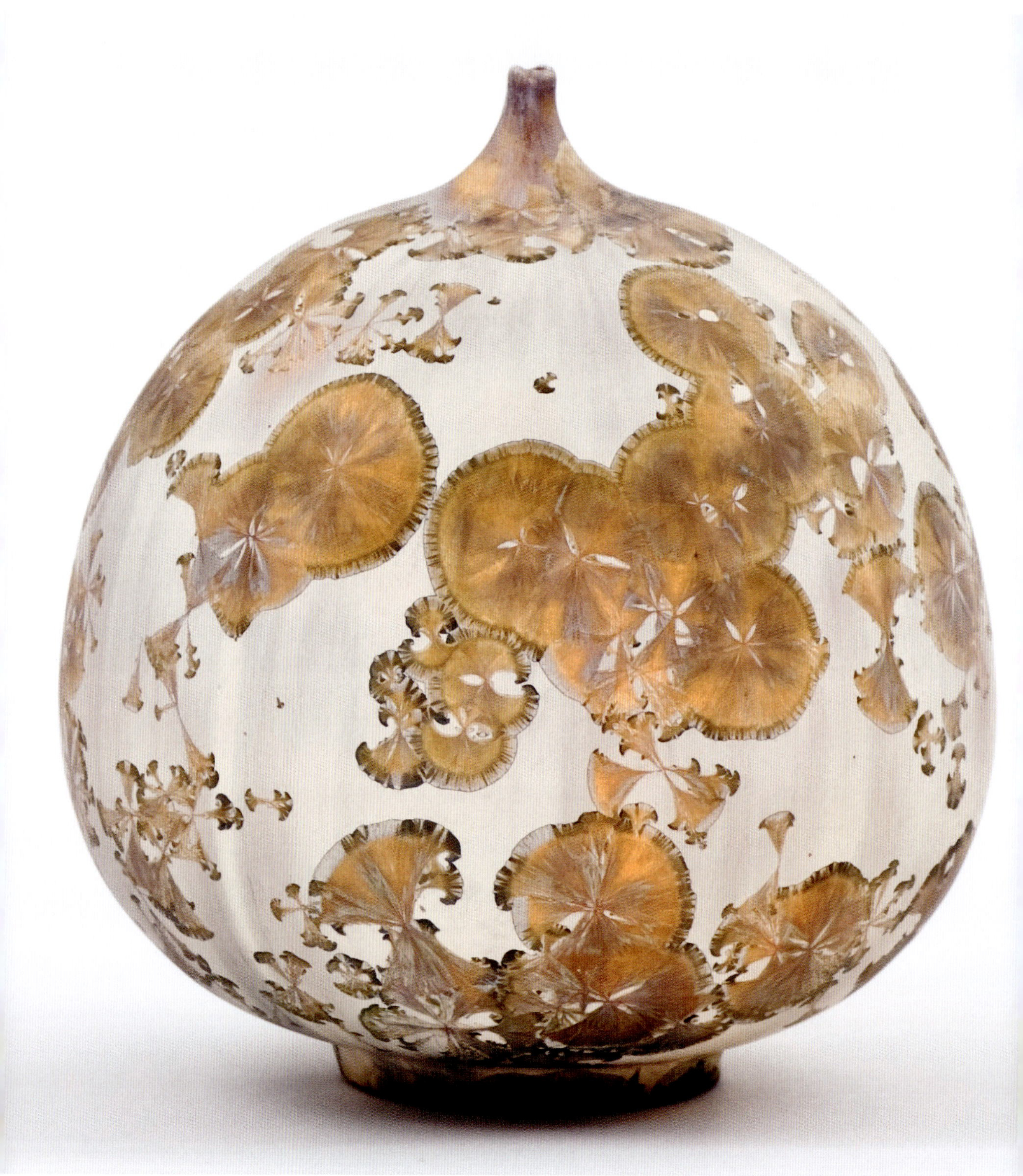

16

Special effects

Once the firing is complete, there is still more that can be done to alter the finished look, and one glaze can continue to give even more interesting outcomes.

Acid etching

After a pot has been through its initial crystalline firing, etching the glaze surface with acid can drastically change its appearance. Glazes with a saturation of colouring oxides work best. Copper, sometimes in combination with other oxides, gives the most interesting results. The acid eats away part of the glaze, usually the background. Several acids can be used, each taking a different time depending on their strength. Lemon juice and vinegar are the slowest, but a longer period can be a benefit.

Vinegar is cheap and readily available. Using 10 per cent vinegar, fill a non-metal container to a depth to cover the pot. Immerse it in the vinegar for several hours, even overnight, checking on the progress. Remove when satisfied and rinse off. The gentler, slower leaching with vinegar allows control of the subtleties in the colour shift.

With the following acids, protective gloves must be worn. Sodium bisulphate's reaction is more active, requiring a shorter time in the acid bath; the same is true of Clorox® or pH Down (used as an additive to swimming pools). Using a stronger acid such as hydrofluoric, hydrochloric, or muriatic acid is not recommended. Hydrofluoric acid is highly toxic and can quickly penetrate the skin and deeper tissue. Hydrochloric acid is a stronger version of muriatic acid, but both are corrosive. They will etch glass and can cause severe skin burns and tissue death.

LEFT Before etching, silver crystals with a gold background. After etching, glossy gold crystals ringed in brown with a pale matt grey background.

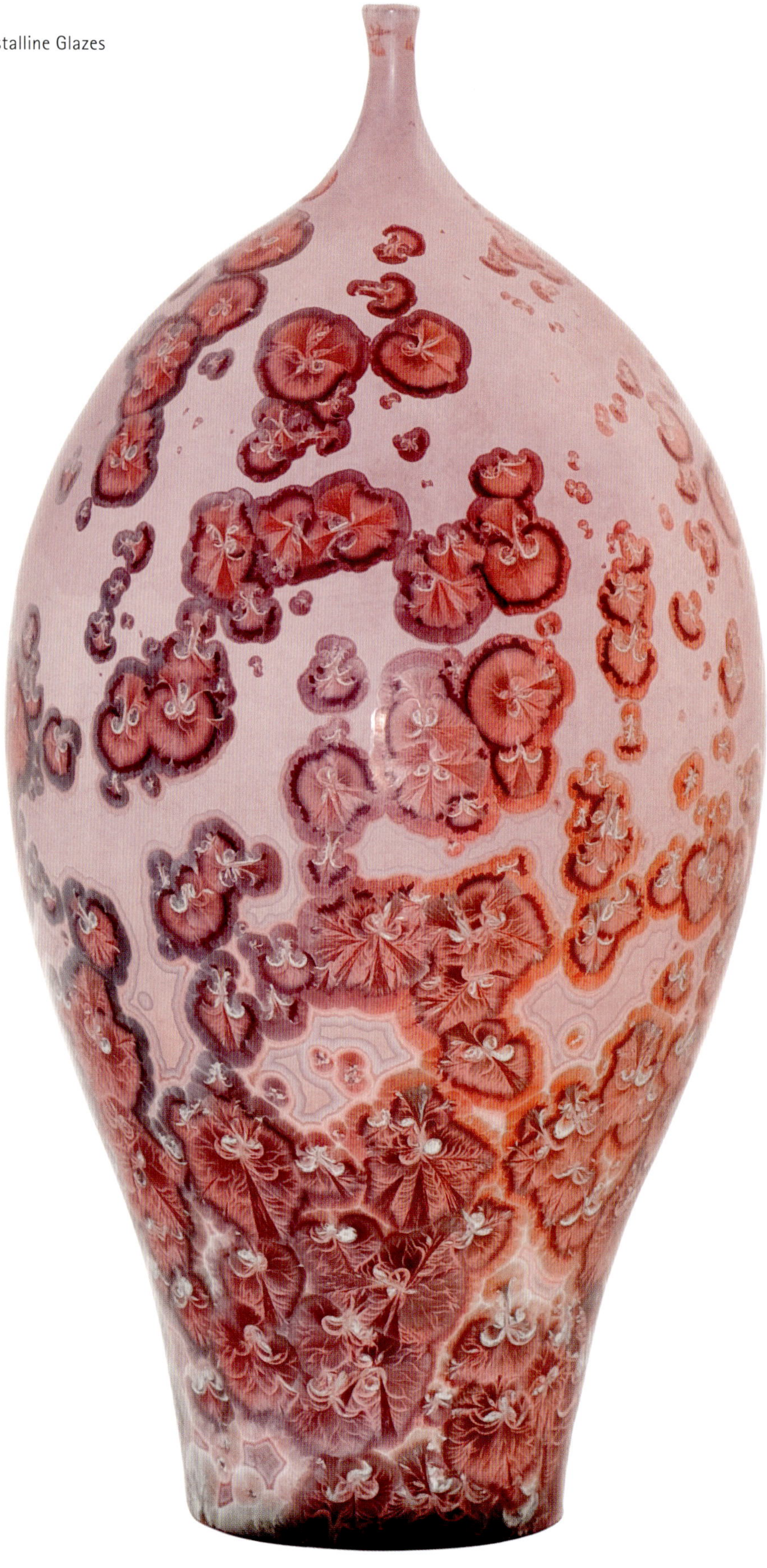

LEFT Reduction-fired copper glaze. Before etching the bottle had glossy dark red crystals with a red background. After etching, the background paled and became matt while the crystals remained glossy.

RIGHT Same glaze as purple bottle on p.117 (Fig. 3), etched in vinegar. Background and crystals have become matt.

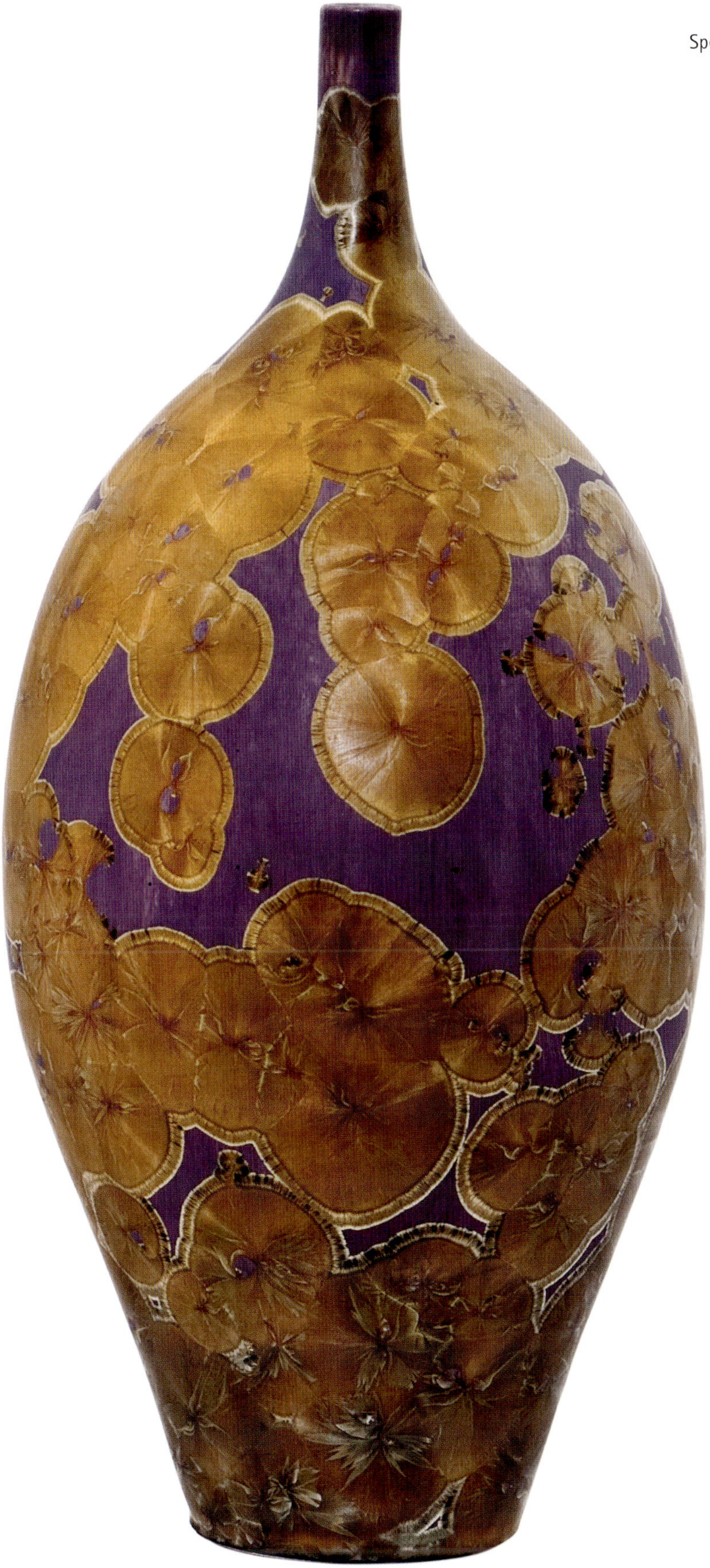

Strike firing

Strike firing is a glass-blowing term that refers to any colour change due to heat treatment. Potters use the term to apply to a firing procedure that involves reheating the piece to achieve different effects. How great the change depends on the colourant and whether the strike firing process is in oxidation or reduction. Some colourants give dramatic results, others hardly any at all.

For an oxidation strike, the finished pots are loaded into the kiln, the kiln is taken to cone 017, then shut off. It's that simple. Glazes with 3–5 per cent titanium dioxide or rutile work best together with iron, cobalt or manganese.

Copper is very sensitive to reduction strike firings. A glaze with as little as 3 per cent copper carbonate may have silver green crystals with a green background. After the strike, it becomes a deep, silvery dark green with a mirror finish. Other colourants also give dramatic results.

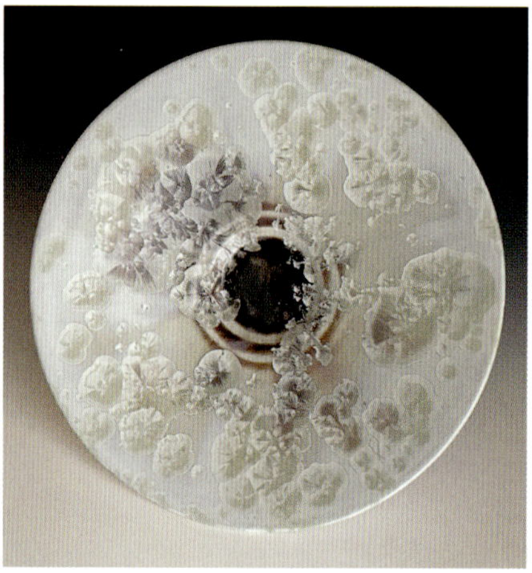

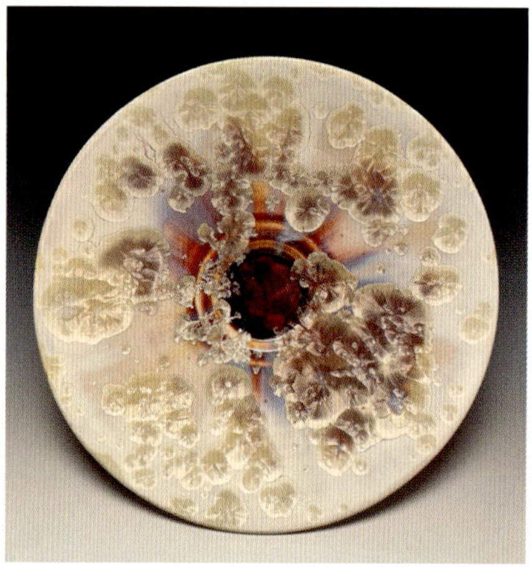

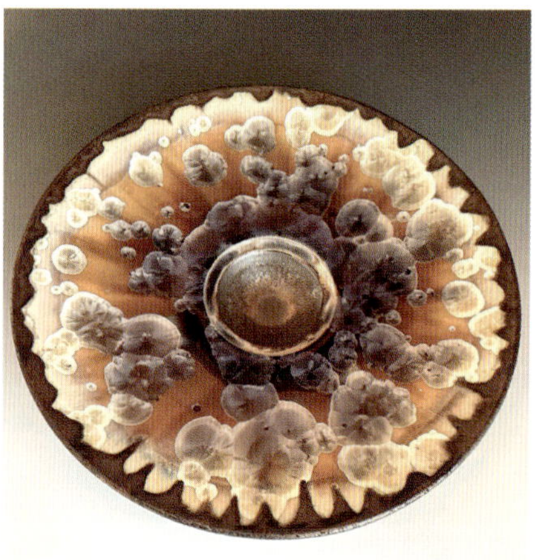

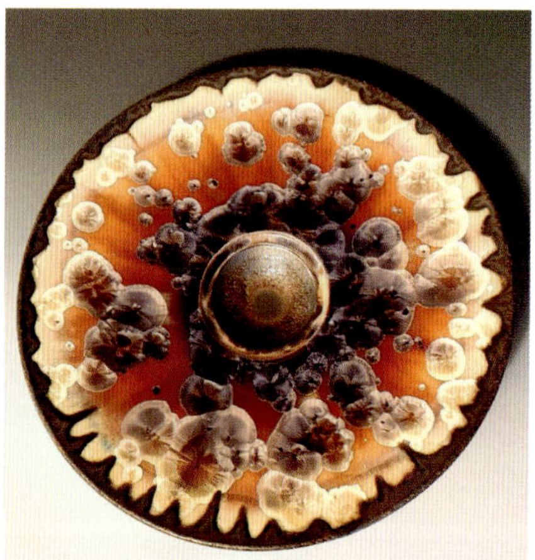

FAR LEFT Plate before strike firing. All images © Holly McKeen.

LEFT Plate after strike firing.

BELOW FAR LEFT Plate before strike firing.

BELOW LEFT Plate after strike firing.

Silver nitrate

Silver nitrate can offer some spectacular results, often creating an iridescent effect as the pot is turned to the light. However, using this compound should only be attempted by those with experience in multiple firings with crystalline glazes, and with a strong knowledge of material handling. When working with silver, it is necessary to wear a respirator, safety goggles and protective gloves. Once it is mixed with water, silver stains everything it touches. If it gets on your skin, it will leave dark brown marks that may remain longer than a week. It will also stain your spray gun and brushes, and your work area should be covered with a protective plastic cloth. Silver nitrate is expensive and can sometimes be difficult to obtain.

Silver nitrate is light sensitive, so only prepare the amount you are going to use at one time. Store any leftovers in an opaque container and keep in a dark place. If it is added to an unfired glaze and left in the light, it will soon turn the glaze a dark grey colour. This has no effect on the final product.

Silver nitrate can be fired in both oxidation and reduction post-strike firings. Amounts needed differ depending on the firing and the results vary greatly.

Silver nitrate in oxidation post-fire strike

Holly McKeen (Canada) creates a wash with one-third silver nitrate to two-thirds hot water. This is brushed over any unfired base crystalline glaze. Results vary depending on the frit used and the amount of silver nitrate in the wash. Silver reacts well on its own, or with other oxides added to the glaze, especially small amounts of copper, cobalt, and iron. The base glaze should have at least 5 per cent titanium dioxide.

After going through a typical crystal firing, the pots are then refired to cone 017 in a strike firing. The results are greater brilliance with reds, blues, and stunning background effects.

Silver nitrate in reduction post-fire strike

Less silver nitrate is needed when used in post-fired reduction. If you use a greater amount, the pots usually go completely silver and may later tarnish.

Base glaze with silver nitrate added: Cone 10 (Jamie Kozlowski)

INGREDIENT	PARTS
Frit 3110	50.00
Zinc oxide	25.00
Silica	25.00
Total	100.00
Add:	
Titanium dioxide	5.00
Silver nitrate	6.00
Cobalt oxide	0.30

After the initial crystalline glaze firing, the pots usually come out looking rather flat. Jamie warns that the glaze is sensitive to anything that touches it after the first firing, so be careful handling when placing the pots in the post-fire kiln, otherwise finger marks will later show.

Jamie does the reduction in an electric kiln, although it can also be done in a gas kiln. Once the temperature reaches 816°C (1500°F), he starts reducing using denatured alcohol in an IV drip.

Jamie's firing schedule

RAMPS	TEMP °C	TEMP °F	HOLD
Ramp up	816	1500	10 min
Ramp down	782	1440	10 min
Ramp down	749	1380	10 min
Ramp down	716	1320	10 min
Ramp down	682	1260	10 min
Ramp down	649	1200	10 min

It is during reduction that the metal oxides within the glaze transform into metallic, iridescent lustres and a myriad of colours can be expected. A single glaze can appear different in each firing, and one glaze may have many distinctive looks depending on which shelf it was on or how close it was to the flame.

Whether the additional strike firing is done in oxidation or reduction, experimenting is necessary to find the best results to suit each kiln firing. Once again, it is the firing that gives the diversity.

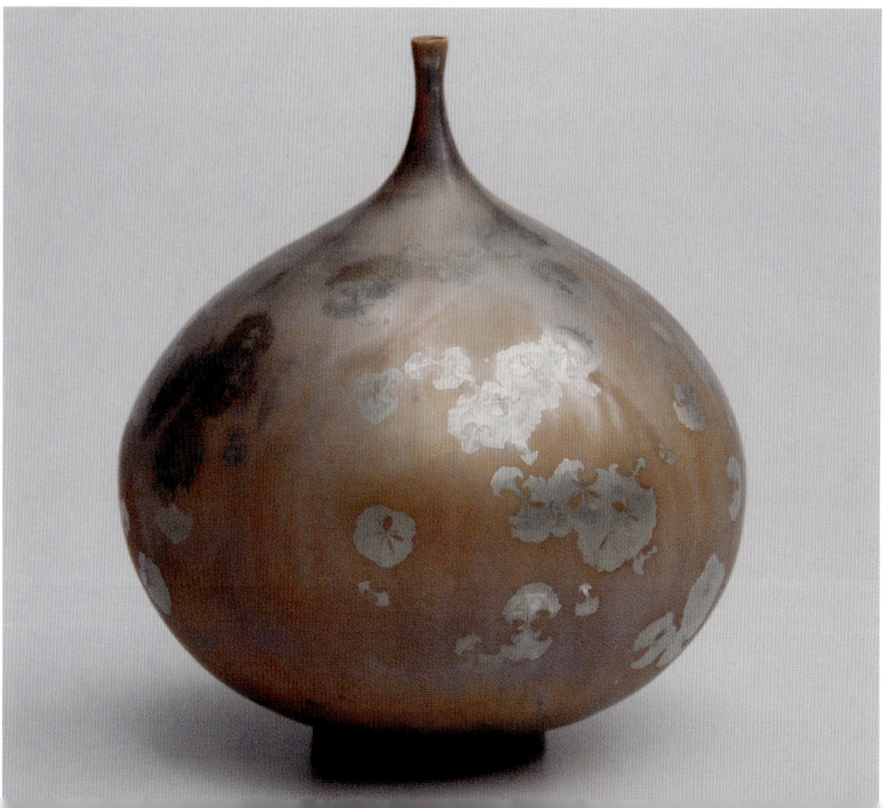

LEFT AND RIGHT Silver nitrate added to glaze then reduction fired.

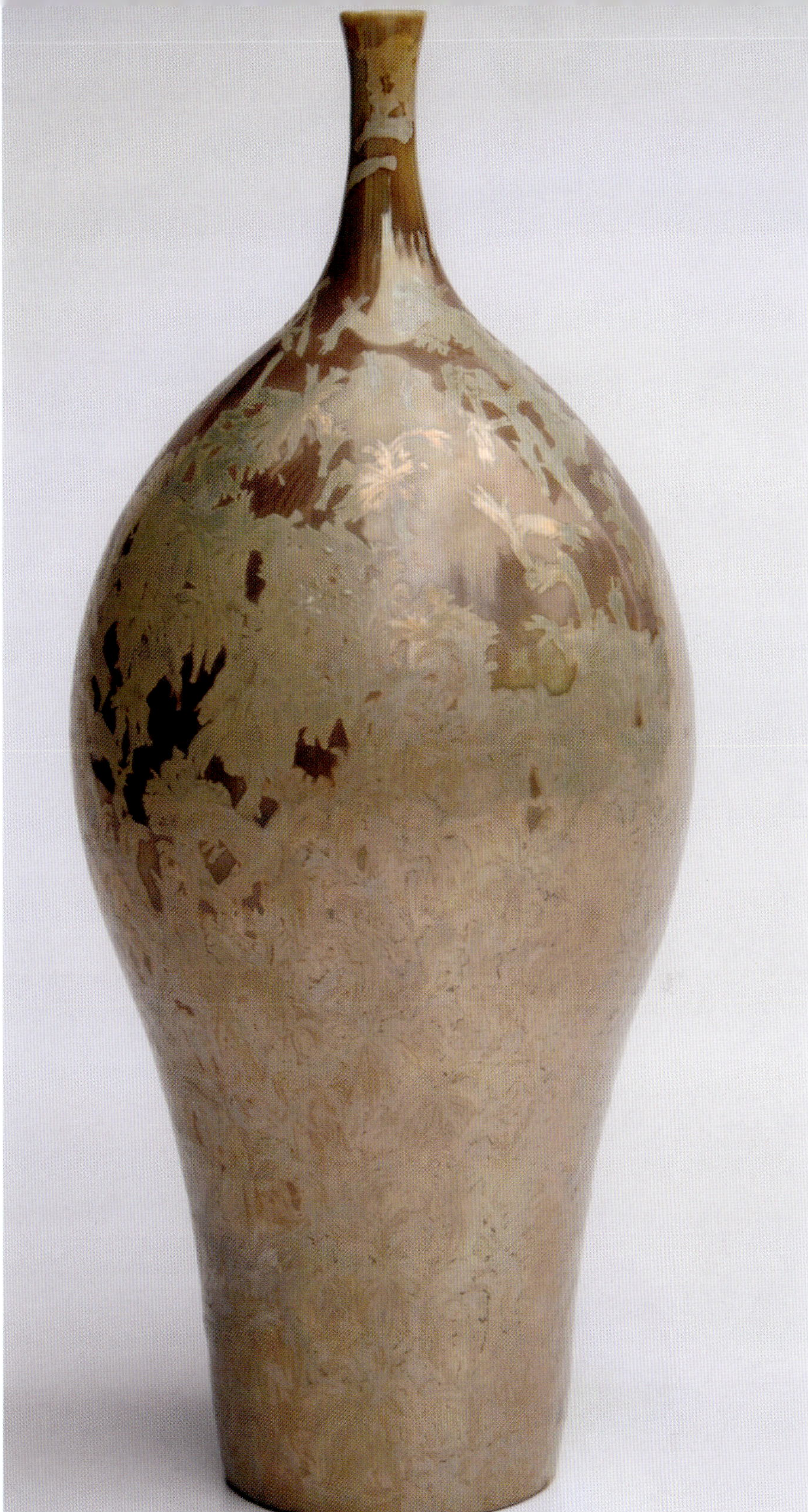

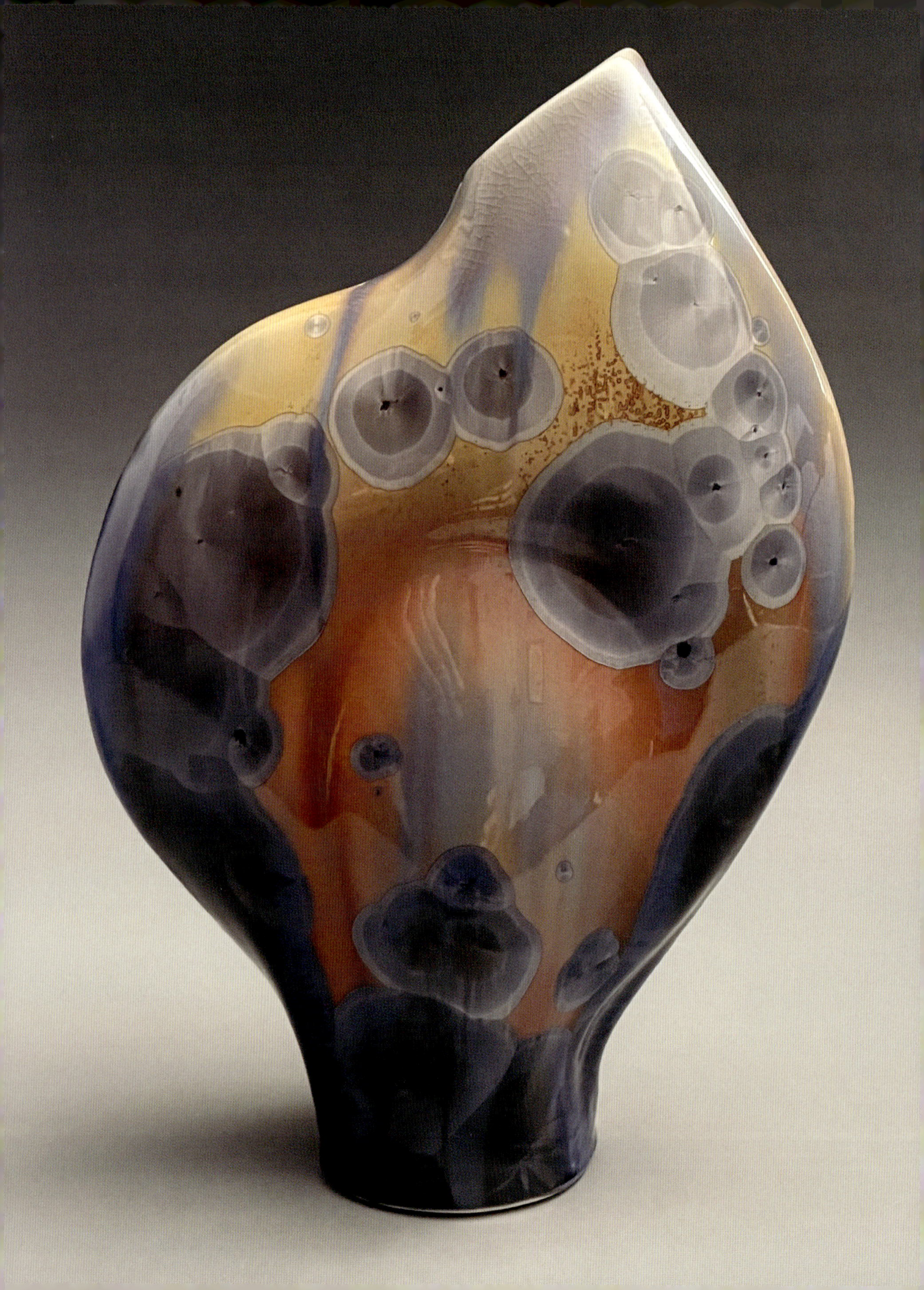

17 Attitudes and aesthetics

LEFT © Ginny Conrow.

BELOW RIGHT
© José MARISCAL Paneque.

OVER PAGE LEFT
© Andrew Krueger.

OVER PAGE RIGHT
© Mariela Tiferes.

Crystalline glazes hold the power to elicit strong emotions. Watch as someone discovers a crystalline-glazed pot. They pick it up with great care, turn it to the light, and may even stroke it. Nevertheless, these glazes have their critics. Because of their ability to attract attention, a crystalline glaze is sometimes thought of as little more than a pretty face or a cheap trick. After the dazzle, what else is there? In *Clay and Glazes for the Potter*, Daniel Rhodes says, 'The presence of spectacular crystals on the sides of a pot, interesting though such crystals may be in themselves, has, in most cases, contributed as little to the esthetic significance of the piece as it has to the function.'

I agree with Rhodes that a unique glaze by itself does not ensure aesthetic significance. Mastering crystalline glazes is only part of the equation. They demand discreet use. Technology must not overpower artistry. Simply because a pot has an unusual or dazzling glaze does not increase its artistic value. Aesthetic significance depends upon harmony between surface and form.

Learning to produce crystalline glazes is no different than learning the science of working with any glaze. It is no more difficult than mastering copper reds, lustres, or salt glazes. But to really know when to use such a finish is both an acquired skill and an expression of aesthetic sensitivity. Most potters strive towards making beautiful forms in harmony with their glazes. To create a surface which suggests frosty windowpanes, rare gemstones, or beautiful flowers is an exhilarating experience.

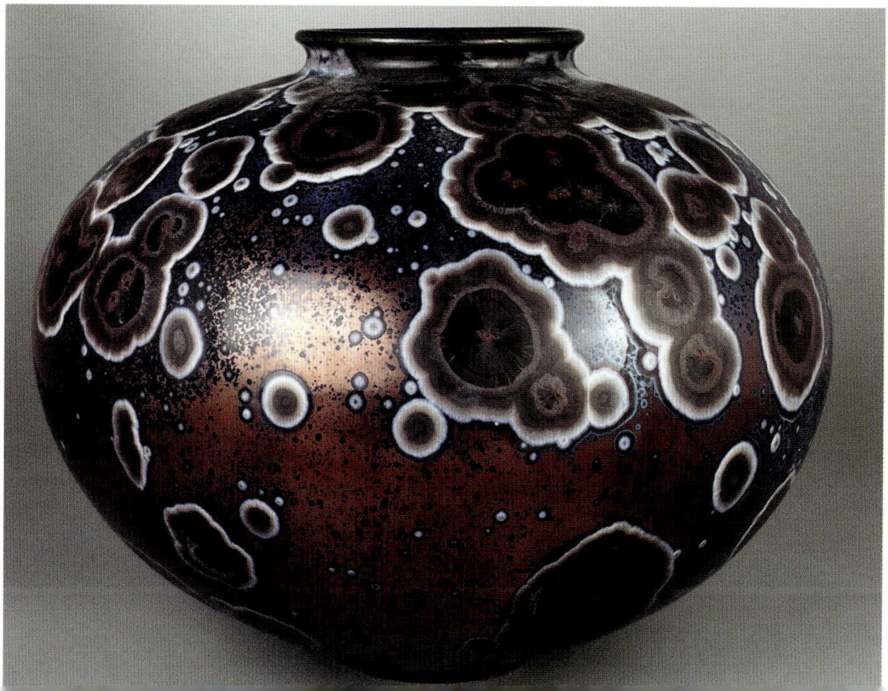

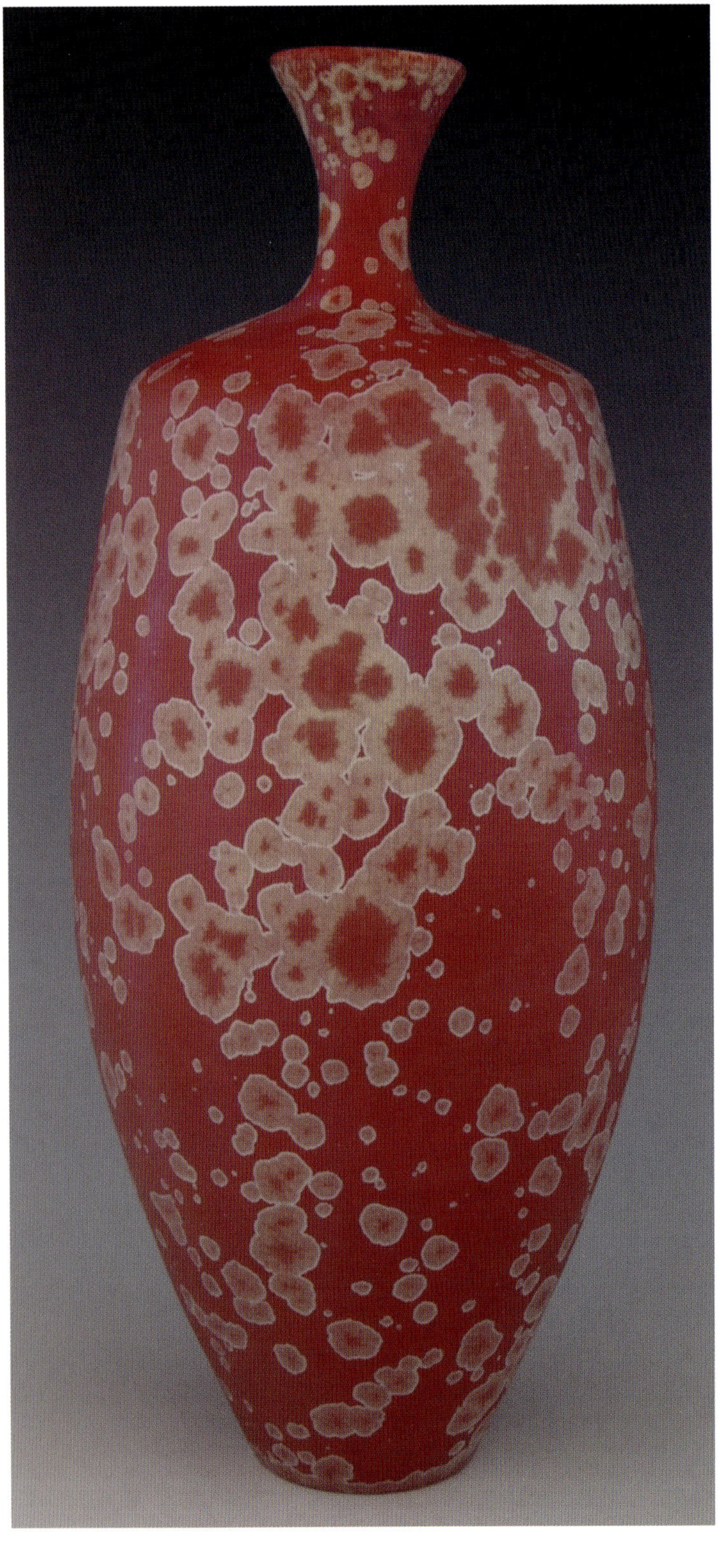

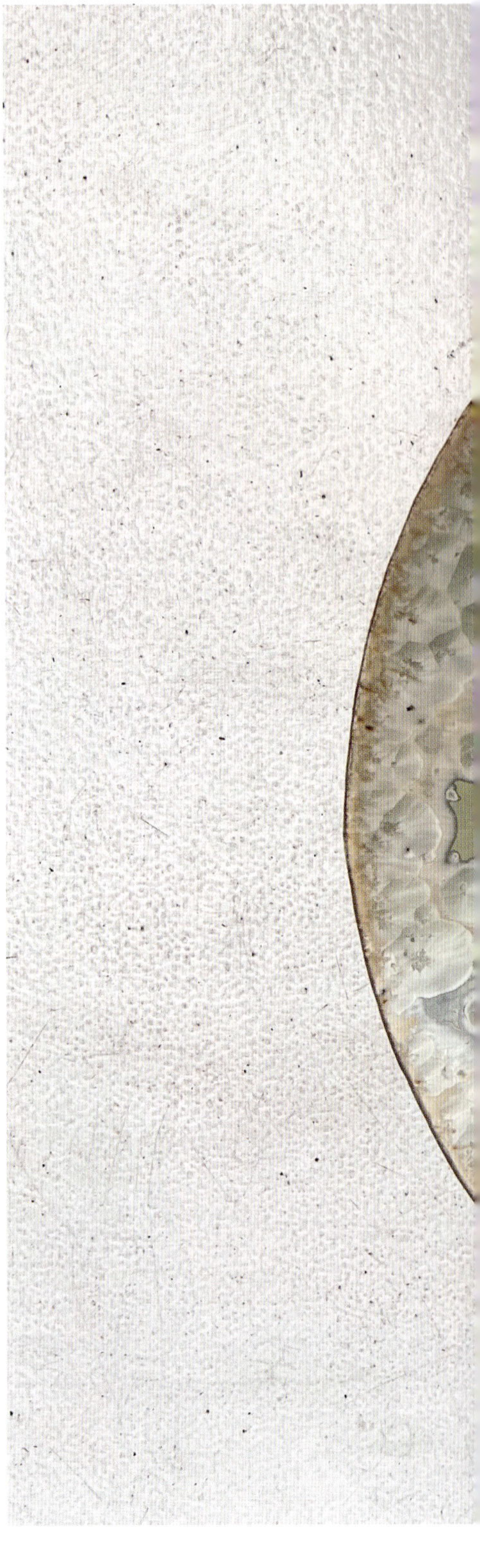

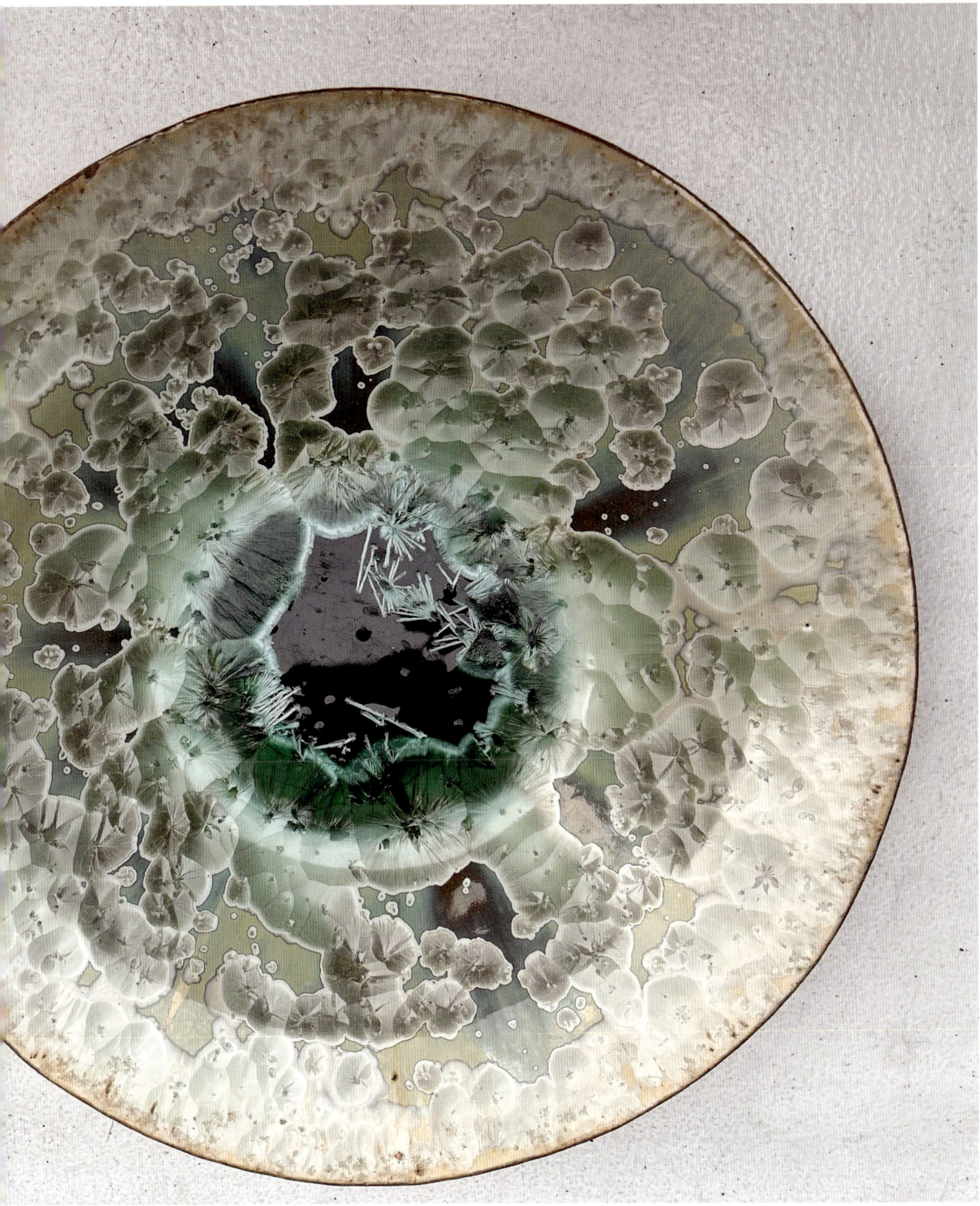

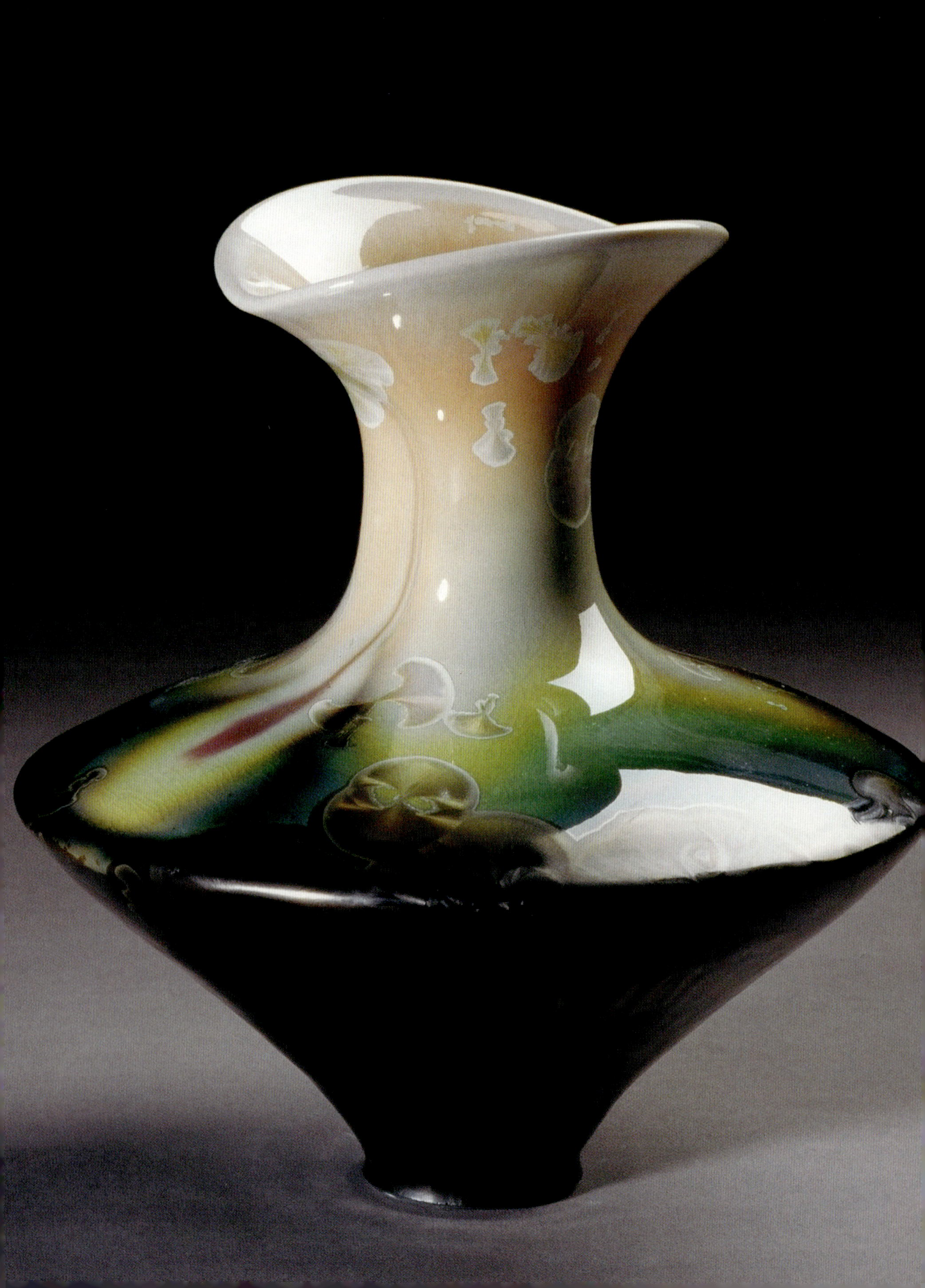

18 Photographing crystalline glazes

LEFT Vase © Ginny Conrow.

High-quality images of a potter's work are prerequisites for submissions to galleries, publications, and competitions, and also for displaying on their website or for online sales using sites such as Instagram and Facebook. Having excellent images is necessary if the work is to be taken seriously. A beautiful pot which is poorly photographed does not attract attention. High-quality photos sell; poor photographic images get passed over.

A gallery owner or potential buyer wants a portrait of the piece. Displaying the pot in a busy background, with someone holding it, or with flowers stuffed inside detracts from the subject itself. The flowers may be admired at the exclusion of the pot. Placing the piece against a plain white background emphasises the true colours of the glaze without reflecting colour onto the pot's surface. If the glaze is light coloured, a black background is suggested.

Crystalline-glazed pottery – or any pottery with a shiny surface – is difficult to photograph. Not only can the photographer, camera, and surrounding objects be reflected on the pot's surface, but you can often have a nasty highlight that detracts from the overall effect. However, there are a few tricks to help you capture the best images possible without hiring a professional photographer.

Lighting

The more control you have over lighting, the more you'll have over glare. Sometimes merely setting the subject out of the sun in a shady spot reduces the glare. Always try for either soft and evenly-diffused light conditions or use natural light by placing the object indoors close to a bright window. Spreading out light sources more evenly may also reduce the amount of glare.

Diffusion can be accomplished in several ways: having more than one light source will spread out the light and create fewer shadows. Photography lamps with high lumens (candle brightness) bulbs have a broader surface than a spot beam and distribute the light over a wider area. Some camera stores allow you to rent studio lighting equipment.

The light can be bounced off a reflector or large piece of white board so it doesn't reflect on the item being shot. To diminish shadows under the pot, a lower light source in front of the object brightens the area.

Some potters have tried to eliminate glare by spraying the pot with hair spray. It does dull the glaze but gives a false impression.

Indoor photography

Dedicating a space in your studio or home for photography can be straightforward. It doesn't have to be large; it can even be a small table. Hang a sheet of photographic paper behind the table, about 1 m (3 ft), and let it fall across the tabletop and drop in front (banquet paper is less expensive and comes on a 1.2 m (4 ft) wide roll). Where the paper meets the table, it should curve at a right angle without leaving a visible crease.

Another option is to buy or build a photo light tent. The one I built is an enclosed cube made of white cloth with an opening in the cloth on the front where the camera lens pokes through. A pot is positioned inside the cube and strong lights are placed in front and on either side of the cube.

The camera should be on a tripod around 0.75 m (2 ft) from the subject to be photographed. Set the camera to the lowest ISO, fast shutter speed, narrow aperture, and shallow depth of field. Using a cable release prevents jiggling, and you don't have to remain crouched behind the camera. Sitting on a stool is more comfortable while lining up the shot. Aim at the belly of the pot and shoot.

Outdoor photography

If you don't have photography lights or your indoor light source is inadequate, shooting outdoors is an alternative. Select an area in the shade with good light, or when the weather is bright but overcast. You can use a table and paper set-up or the light box with the light source (sun) shining on the item.

Before shooting, look through the lens to see if there is any reflection on the surface of the object. Do you see yourself? Trees? Clouds? If you see yourself, crouch out of range and take the photo with the cable release. I have even placed a bedsheet over myself and the camera, with a small opening for the lens. If the background is a problem, hang a sheet behind the pot. Take several shots at different speeds.

You can also take images with your phone. Some phones cameras are even superior to digital cameras. Tripods are also available for phones.

Experiment by taking several shots in a variety of conditions. If you have problems and aren't satisfied with your results, talk to a consultant at a good photography store. They are usually interested and can often offer positive suggestions.

Adobe Photoshop can be used to crop the image, tweak colours, and adjust lighting. Corel PaintShop Pro is a reasonable alternative to avoid paying a monthly fee for the more comprehensive Photoshop software. Other software is also available, e.g., Adobe Express, Gimp and Acorn.

TOP RIGHT Large plate © Andrew Krueger.

BOTTOM RIGHT Bottle © Bill Boyd.

RIGHT © William Schran.

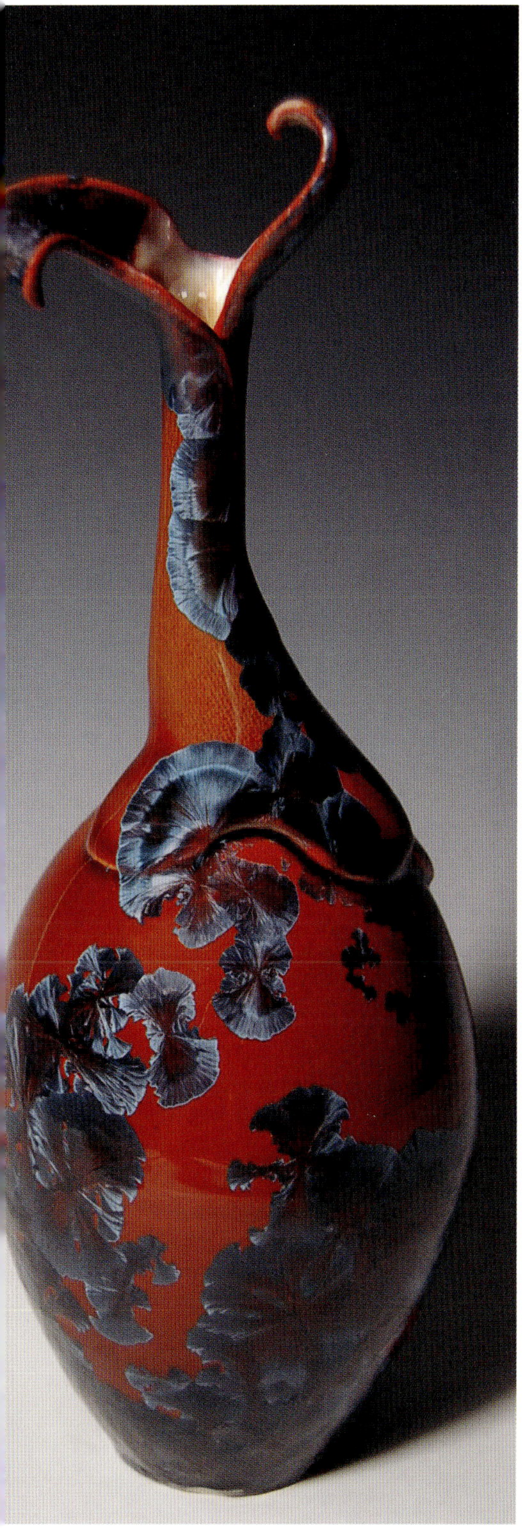

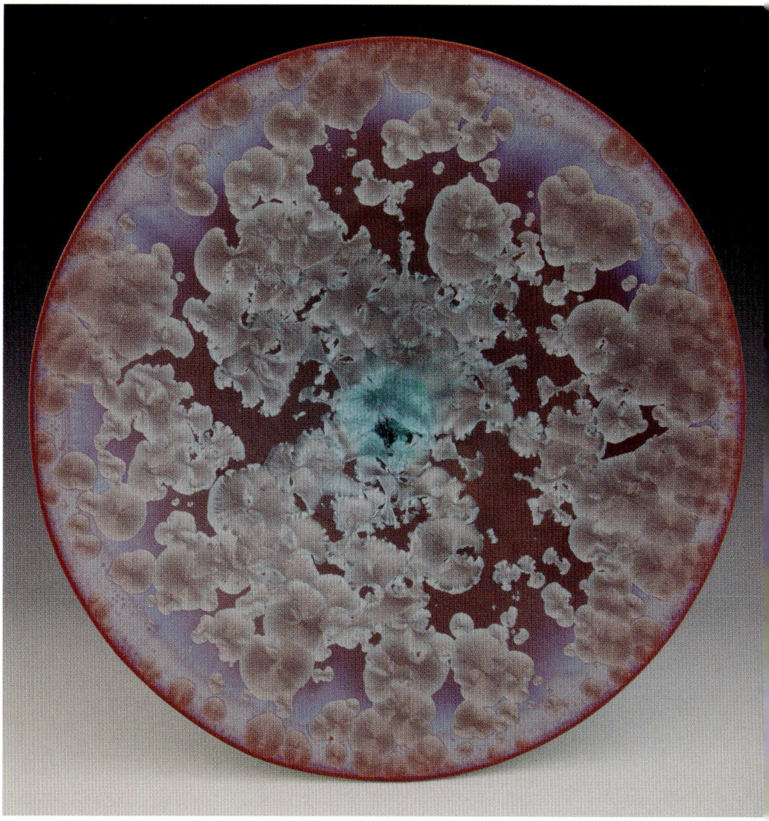

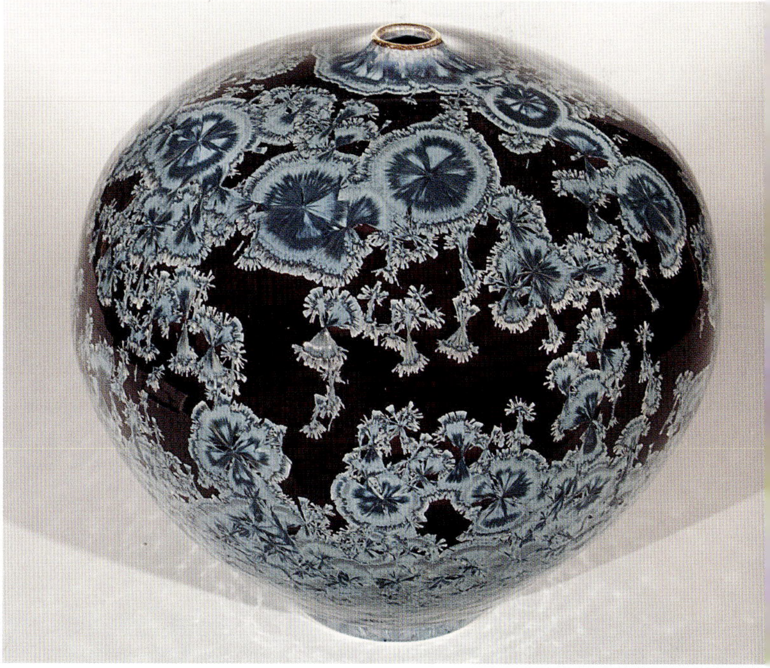

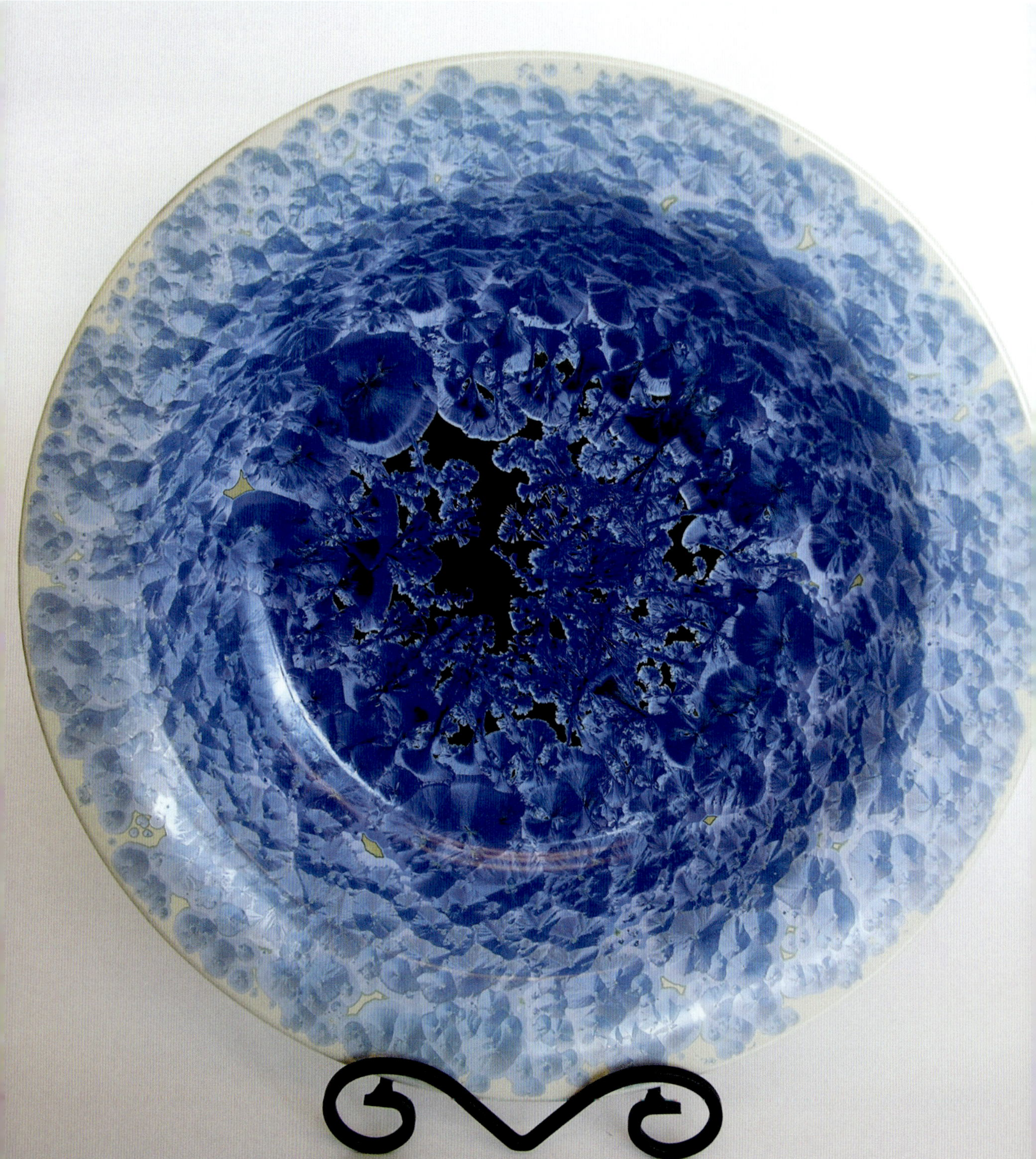

19 Marketing crystalline glazes

LEFT Large plate, 24 inches across, fired in oxidation.

Considering all the hard work and expense that goes into producing crystalline glazes, it is important to have a market that allows you to at least recoup your costs and, if this is your profession, earn a living. Belonging to a craft guild and selling at seasonal craft sales are outlets. Some potters also sell from their studios. Both these situations can be problematic.

Pottery was my full-time profession. I trained as a production potter and worked only in porcelain, making dinnerware sets, teapots, mugs, bowls, plates, and vases. There was always a market for these items. When I switched to crystalline glazes, suddenly my income took a nosedive. I needed to find a way to continue to earn an income and still make the pottery I loved the most.

I worked alongside my husband, a prolific stoneware production potter. We had a large display area at the front of our studio. My crystalline pots placed next to his work didn't sell; my bottle was ten times the price of his coffee mug of the same size. At craft shows, his pots flew off the shelf, while my crystalline glazes were admired but hardly sold. I quickly realised we couldn't sell together. My work was considered too expensive in comparison. When I got my own display booth, my sales picked up.

We created a separate area just for crystals in our showroom, with glass shelving and brighter lighting. If I had several pots that were similar, I only displayed one at a time, making each pot unique. Information was posted that explained the crystalline process. By educating the customer in the extra labour involved, my work got more attention and sales increased, but they were still slow.

My next move was to see who was buying my work. I realised there was a market selling in high-end craft stores and art galleries. The customers who frequent these places are looking for – and willing to pay for – art. Granted, the gallery owner takes 50 per cent of the sale price, but I was making the same income as before, and spending less time on promotion and selling. I did find, however, being accepted into galleries and craft stores usually required excellent photos as well as an up-to-date and interesting website.

The extra work of making catch basins and the more complicated kiln firings meant I needed to find a way of working with crystals that was also cost efficient.

Over time, I found that working by the seasons was my best approach. I started the new year making hundreds of small bottles and bowls by throwing off the hump. These are quick to produce and became my glaze tests. If they turned out well, I sold them. Fortunately, they were popular items. I would get $80–100 for a little bottle or bowl with a nice crystalline glaze and could throw them faster than a $21 coffee mug, which required a handle. I would make several thousand of these small items before I progressed to the larger, one-of-a-kind pieces. These were thrown on batts, and were usually vases, large plates, bottles, and covered jars. By now, several loads of the test pots would have gone through the kiln, allowing me to test the glazes and see the variances in firings.

Once October arrived, however, the crystals were put aside, and I focused on production ware for the holiday season. Working in this manner allowed me to make a decent living and to do what I loved most – crystals – for ten months of the year.

Making crystalline glazes is a passion. Why else would we go to all that extra work? But it is also necessary to get our pots out into the world for others to see and admire. A beautiful form in combination with a mesmerizing glaze is what defines crystalline glazes.

BELOW Oxidation fired bottle with matt glaze.

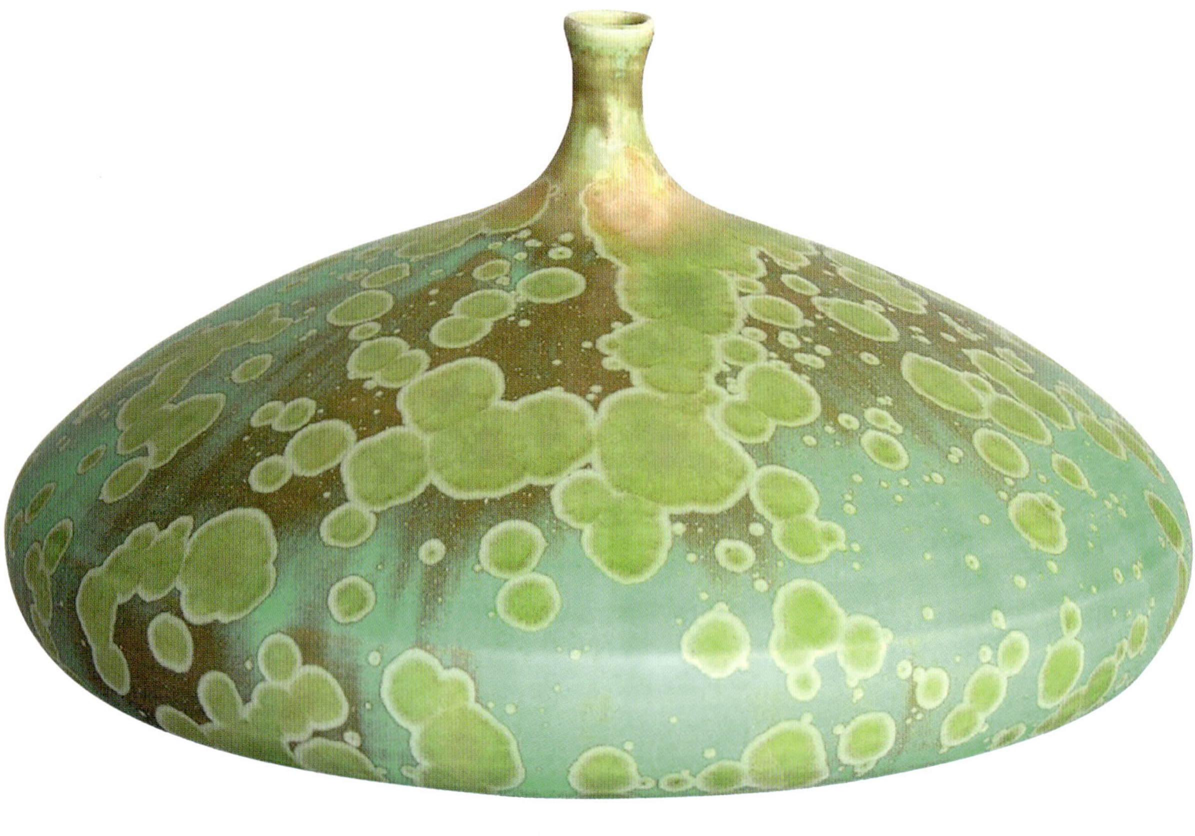

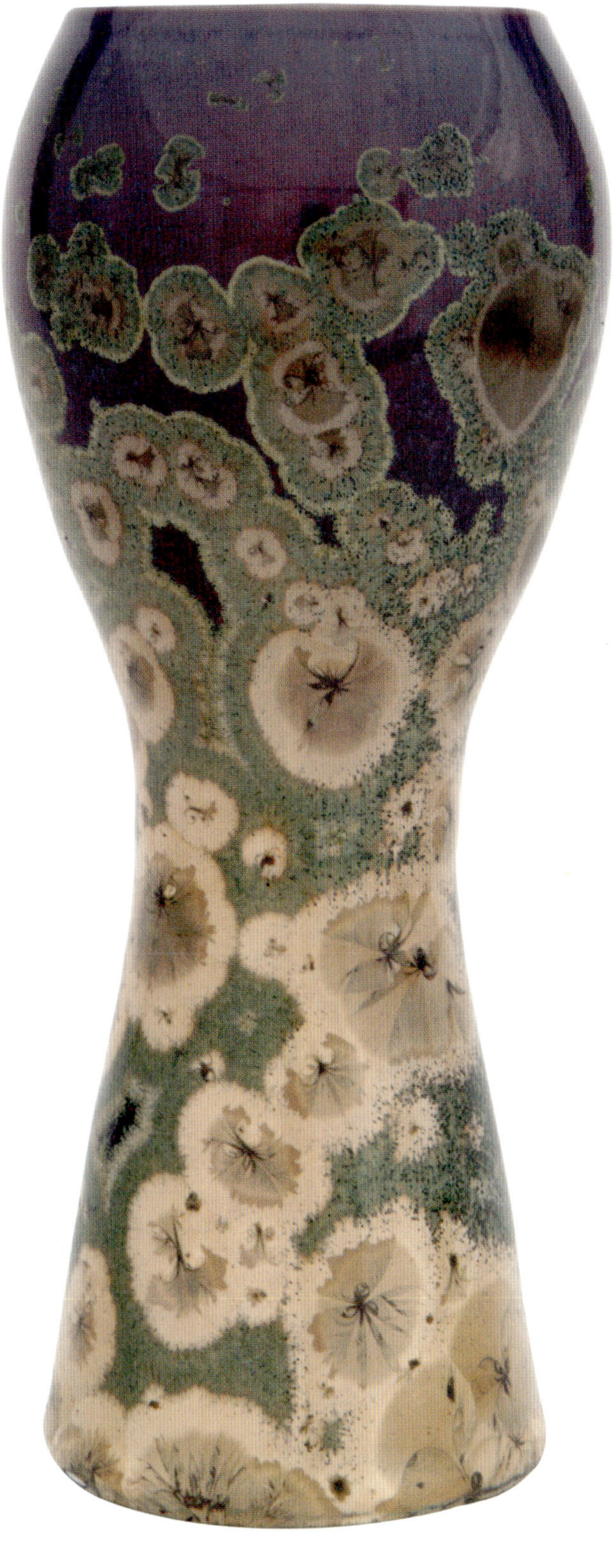

RIGHT Tall vase, reduction-fired.

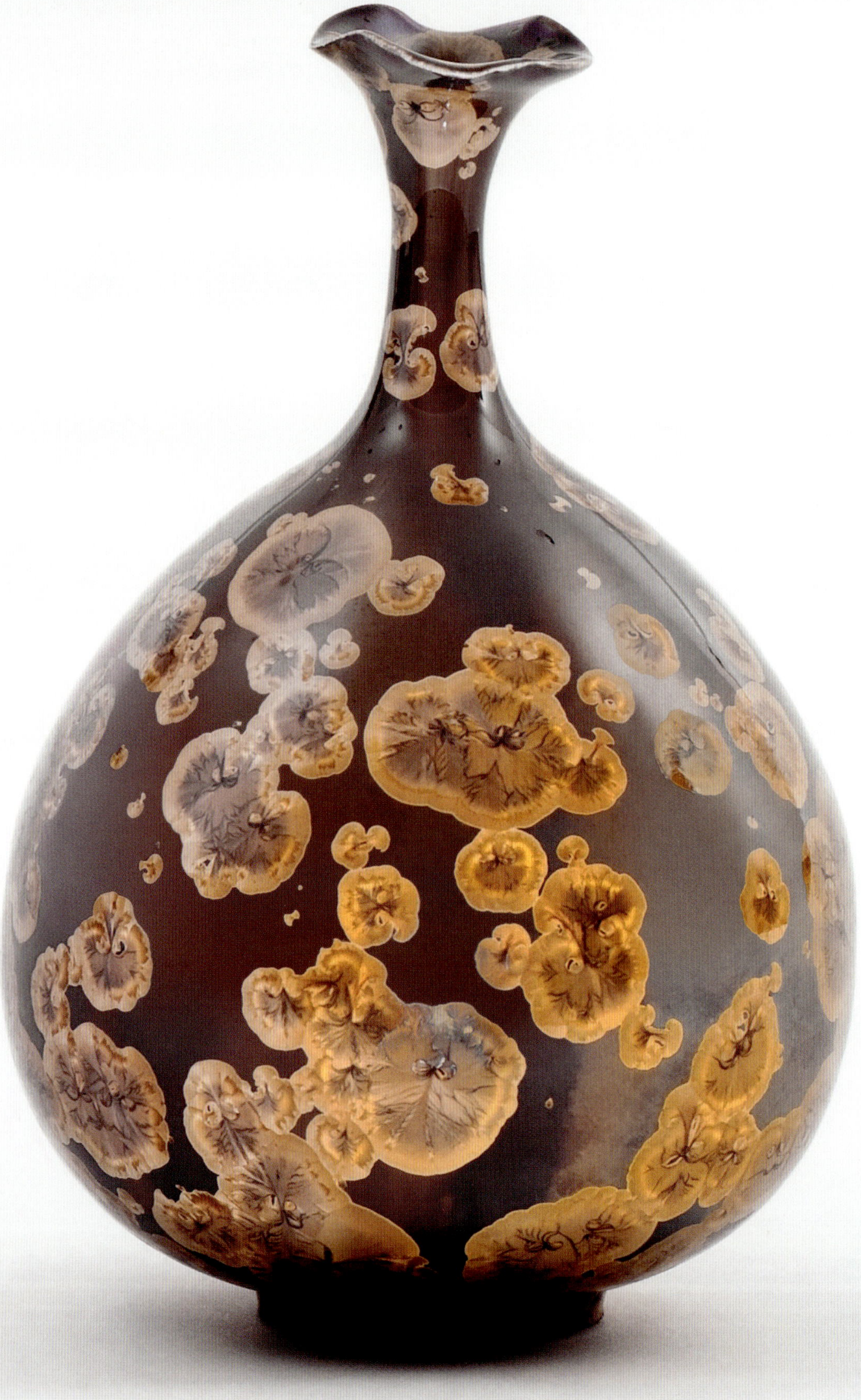

LEFT Manganese crystalline glaze fired in oxidation.

RIGHT Tall bottle, matt crystalline glaze with copper and cobalt carbonate fired in oxidation.

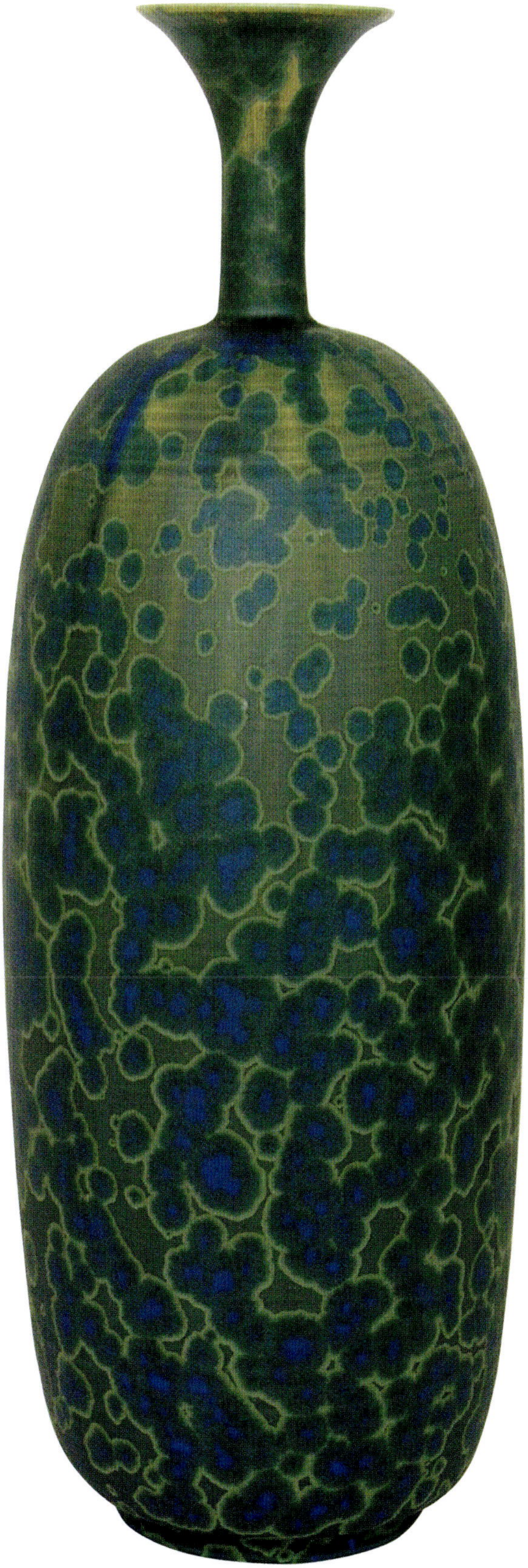

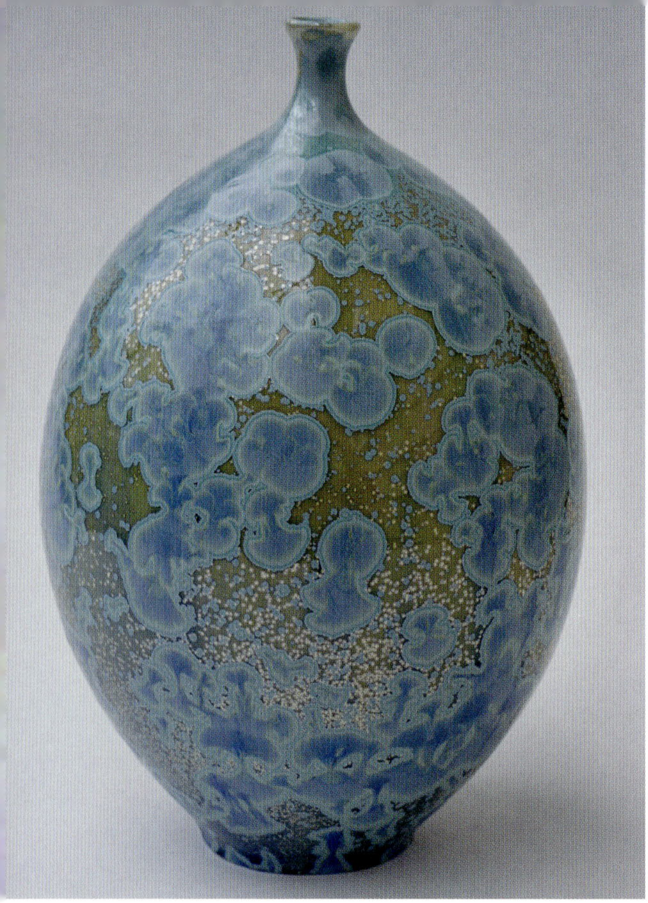

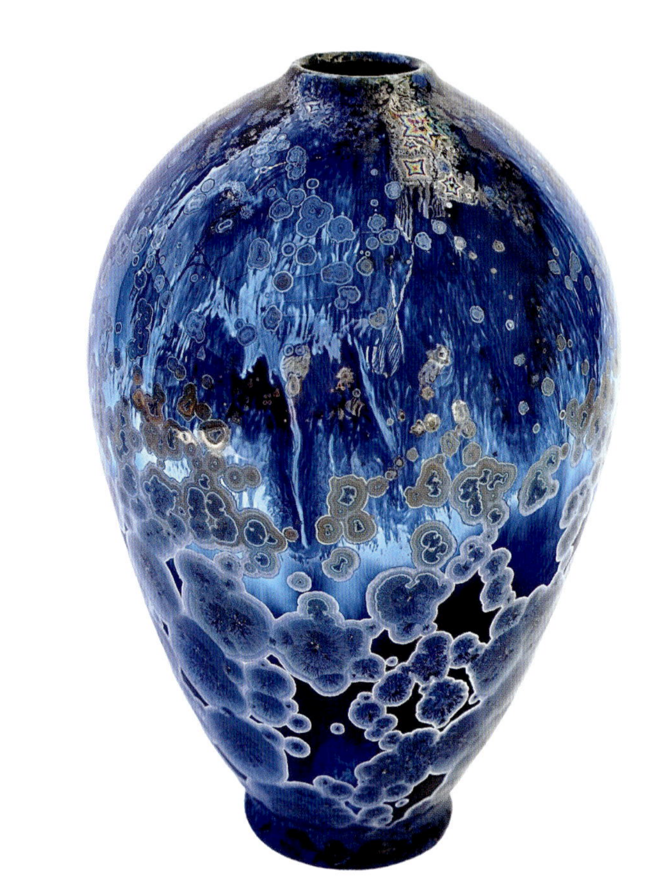

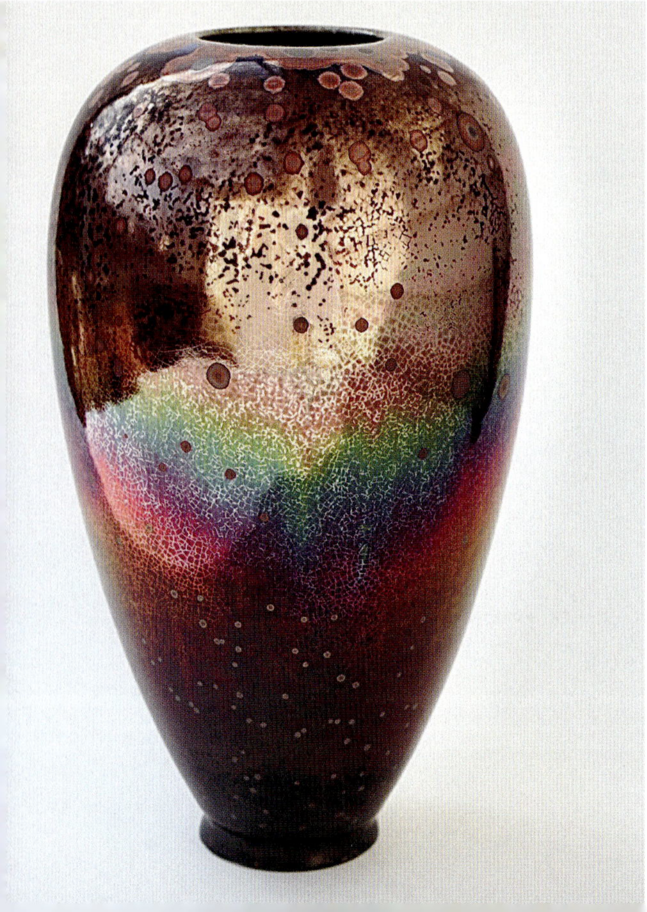

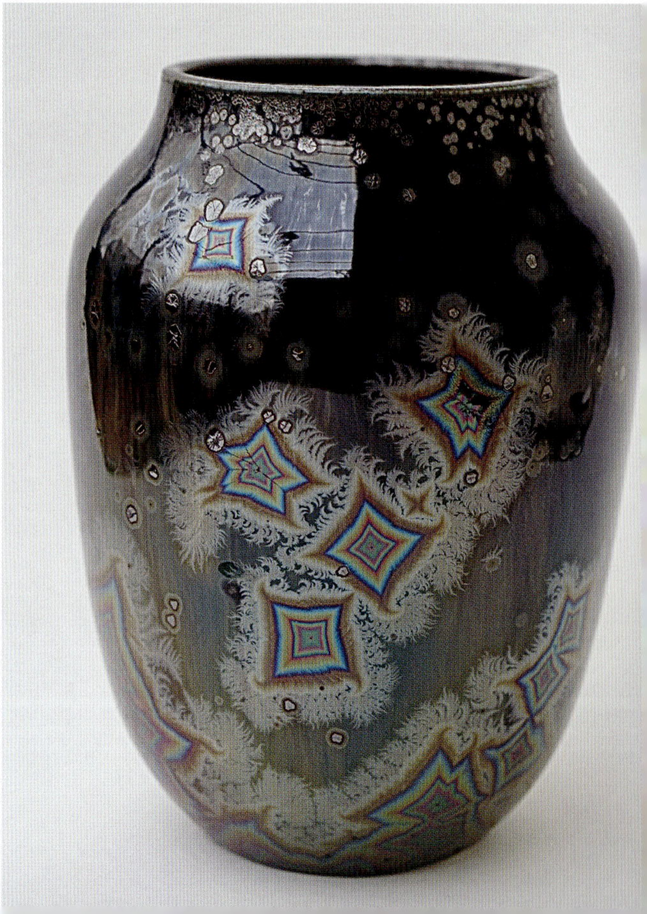

20 Featured potters

LEFT All © Casey Anderson.

Casey Anderson, USA

Casey Anderson made pottery in high school but set it aside after graduation. He didn't take it up again until 12 years later in 2017 when his wife, Sara, gave him a wheel for Christmas. His apartment living room became his first studio. 'Being a full-time nurse is a demanding job,' he explained. 'Pottery became an outlet to help with the stresses.'

Casey has since converted his garage in Mebane, North Carolina, into a pottery studio. Today he has three kilns: two electric and a converted downdraft gas kiln for reduction firings. But his only wheel is still the one that Sara gave him. Casey says:

> *'In late 2020, I came across a crystalline vase image online and was intrigued. The colours were vibrant, and the crystal structure gave so much depth to the glaze. I told myself that this is what I wanted to create.'*

Casey took online classes, purchased books, read articles online, and took crystalline glaze workshops to further his knowledge. Images of his pottery on Instagram and Pinterest soon generated interest. A couple of years ago, Casey started marketing his work on social media, and at art festivals and exhibitions.

Since the first objects he made were small kettles or teapots, The Dusty Kettle seemed a fitting name to go by when showing or selling his work. Crystalline glazes have now become his passion, and his aim is to make the best work possible. Casey explains:

> *'I'm critical of myself and am always trying to learn and grow as an artist. Once my crystalline journey began, I ventured more into understanding the complexities that come with the art. The smallest variations would totally change the outcome. There is so much more to these glazes than just applying a glaze and firing it. You need to factor in glaze materials, thickness of application, heat work, firing cycles, age of kiln elements, and density of kiln load for every firing.'*

Casey most enjoys these fine details, and the process required to make something magical.

Most of his crystalline glazes require cone 10 porcelain. His favourite glaze is called Rainbow Skies.

Rainbow Skies: Cone 10

INGREDIENT	PARTS
Frit 3110	48.00
Silica	25.50
Zinc oxide	26.30
EPK	1.50
Titanium dioxide	1.20
Copper carbonate	3.20
Cobalt carbonate	1.50
Bismuth	1.50
Total	108.70

Firing schedule (electric kiln oxidation)

RAMPS	TEMP °C	TEMP °F	HOLD
250°C (482°F)/hr	1180	2156	0
80°C (176°F)/hr	1270	2318	10 min
Ramp down	1060	1940	1 hr
Ramp down	1020	1868	30 min
Ramp down	850	1562	30 min
Ramp down	800	1472	20 min

Post firing reduction (gas kiln): two to three hours to 600°C (1112°F) in oxidation. Next, reduce while raising the temperature slowly, from 600°C up to 720°C (1112–1328°F). This takes about an hour, after which the kiln is shut off.

> *'Every time I open the kiln, I don't know how it will have turned out, which makes each firing an emotional experience, whether good or bad. Crystalline glazes have infinite possibilities, and it is a never-ending journey. This is what drives me to continue on this path.'*

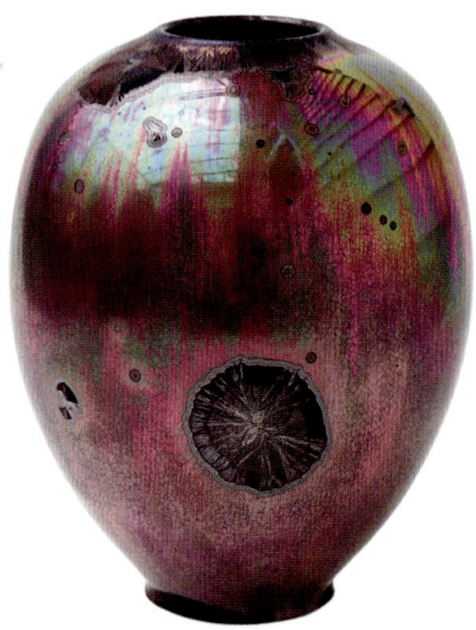

LEFT © Casey Anderson.

Richard Bideau, England

Richard Bideau's custom-built studio is close to his Burnley, Lancashire home, in the north of England. His first encounter with crystalline glazes came at the age of 24, in the form of an article published in a 1995 science magazine. Richard explains:

> *'I was working as an engineer at the time, but with a connection to clay already developed through my mother, a long-established potter. One sentence in particular grabbed my attention. It said something like, "Few potters are willing to persevere with a technique which is renowned for being frustratingly unpredictable".'*

That aroused his curiosity and his journey with crystalline glazes began. Starting with a recipe from Derek Clarkson, a well-known British potter, Richard has experimented and developed thousands of glaze variants over nearly 30 years. 'At the onset, I scarcely imagined the immense possibilities that this obscure branch of silicate chemistry offered. Perseverance has certainly been necessary while exploring what is surely only a small part of the crystalline landscape.'

Richard's preferred clay is porcelain. For many years he used the popular Audrey Blackman body made by Valentine Clays at Stoke-on-Trent. Now, however, he has moved towards a formulation of his own that gives especially good results with his glazes, and stands up on the wheel better than his previous clay.

Over the years, he became frustrated with the variability of frits, especially Ferro Frit 3110. He solved the problem by making his own.

His favourite glaze has many variants achieved through small changes to the chemistry, most of which produce an abundance of different secondary crystals, depending on the firing cycle and porcelain body composition. This glaze also works well in combination with his iron silicate aventurine glazes. While it cannot be duplicated exactly with commercial frits, the following is a close approximation.

Emerald Green: Cone 10

INGREDIENT	PARTS
Frit 3110	24.00
Frit 3195	3.50
Calcium carbonate	2.60
Barium carbonate	8.20
Zinc oxide	28.40
Kaolin (Grolleg)	3.20
Silica	30.10
Total	100.00
Add:	
Titanium dioxide	1%
Red iron oxide	4%
Cobalt oxide	7%

Varying the amount of cobalt oxide changes the colour from pale honey (no cobalt), through to very delicate green crystals on a honey background (0.05% cobalt), through to blue crystals on a blue ground (0.7% cobalt).

Richard designed his own 200 litre (7 cu ft) front-loading electric kiln and its controller. They are specifically built for the demands that the crystalline process places on the firing. There are three independent zones within the kiln, each with a thermocouple and element bank. They are individually controlled to achieve temperature uniformity throughout the firing chamber. The controller has the capability of an almost unlimited number of firing segments, allowing for complex temperature profiles. The vents that let air into the kiln for cooling are motorised and linked to the controller.

Firing schedule

STEPS	RAMPS	TEMP °C	TEMP °F	HOLD
1	200°C (360°F)/hr	1185	2165	0
2	60°C (108°F)/hr	1285	2345	5 min
3	Ramp down	1050	1922	0
4	Ramp up	1080	1976	0
Repeat steps 3 and 4 another four times.				
5	Ramp up	1140	2084	2 hr
6	Ramp down	1000	1832	30 min

Richard sells his work at exhibitions and galleries in the UK.

> *'Gradually, I've moved away from the more vivid glazes and settled on a style that adds the texture of secondary crystals to the usual zinc silicate primaries. My aim now is to capture some of the features of nature: lichen on an old wall, minerals glittering in rocks, flowers in a sun-dappled woodland. When photographing my work, I choose backdrops of the local landscape. After all, crystals are a natural phenomenon.'*

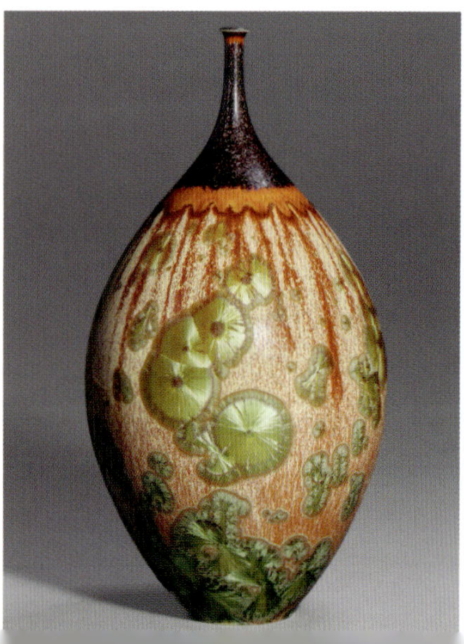

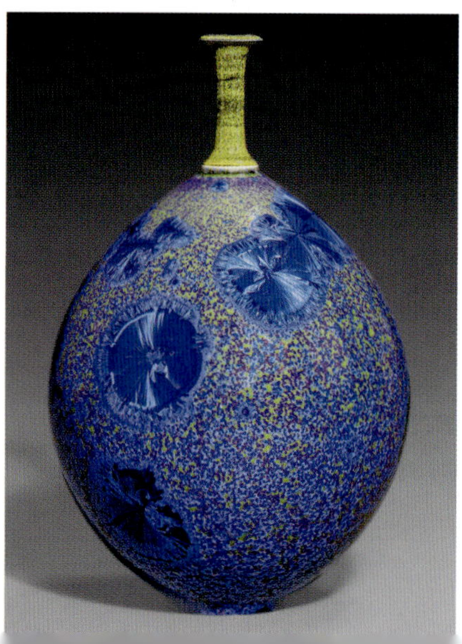

LEFT Both © Richard Bideau.

Bill Boyd, Canada

Bill Boyd initially started selling his work in a tiny showroom at the end of his driveway on Galiano Island in British Columbia. So he wouldn't be constantly interrupted while working, his customers were on the honour system. They would write up their own sales slip, leave the money in a box, and wrap up what they had purchased. This worked well for many years, but in those days he was making inexpensive utilitarian stoneware and selling to a local market. Bill explains:

> *'I don't think of myself as an artist, I'm okay being called a craftsperson who sometimes gets lucky and creates art. And that explains a little about my thoughts concerning crystalline glazes. Growing crystals is an amazing natural phenomenon that I try to facilitate in a way that leads to good results on pottery.'*

When Bill chose to live on Galiano Island, he wasn't thinking of making a living as a potter. He moved there to be surrounded by the ocean, for the profound beauty, quiet lifestyle, lack of traffic, and maritime atmosphere. Bill says:

> *'But living on an island also has its drawbacks. I need to take a ferry to Victoria to deliver my work and two ferries to Vancouver for clay supplies. A small boat used to pick up my pottery, but I once saw the boxes being thrown from the dock into the boat. I immediately stopped shipping for fear of breakage.'*

Now Bill only makes crystalline-glazed porcelain, which he sells through galleries across Canada. He used to sell in the US too, but shipping across the US border involved customs officers opening boxes and often failing to replace the items properly in their containers, which caused considerable breakage. Insuring ceramics for shipment to the US was also expensive, so he decided to limit orders to within Canada.

Allowing galleries to sell his work gives Bill more time to create. The owners have a broader clientele than island residents and know how much their customers are willing to pay. Even though galleries take a 50 per cent markup, they sell far more than he ever could on Galiano, and it wasn't long before his prices tripled.

Bill's studio is an addition to his home. He has two potter's wheels and three Skutt electric kilns that he uses for bisque firing and electric-fired oxidation glazes. He does the reduction firings in a Geil, 0.45 m3 (16 cu ft), fully programmable, propane fibre kiln, usually firing to cone 10/11. He uses approximately ten glazes, but the following is one of his favourites.

Mandarin Night: Cone 10/11

INGREDIENT	PARTS
Fusion Frit F-413	65.00
Zinc oxide	22.00
Silica	9.00
Ilmenite	5.00
Red iron oxide	1.00
Bentone® EW (a highly processed form of bentonite)	0.36
Total	102.36

The Geil kiln allows Bill to pre-set an automated firing programme so he can walk away and rely on the kiln to manage everything. He sets the controller to first fire in an oxidation atmosphere to maximum temperature, 1282°C (2340°F), drop the temperature to the crystal-growing range, then do five ramps by lowering and raising the temperature during a three-hour period. During the fourth ramp at 1093°C (2000°F), he does a 20-minute reduction. Then he does a last ramp to 1010°C (1850°F) and holds in oxidation for 15 minutes. Then the kiln is shut off.

Below is the schedule for firing crystals in reduction in the Geil kiln:

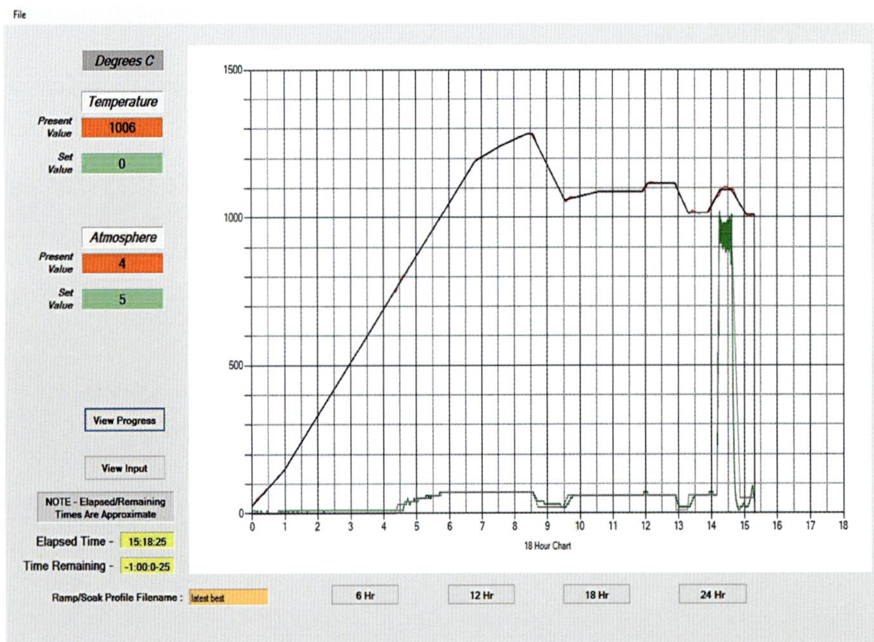

'I often apply gold metallic trim at the top or around the outer edge of the pots and do a third firing. First bisque, then glaze, then gold. I made some interesting discoveries that most potters wouldn't know about because they probably don't

use gold metallics. The pots come out of the gas firing a light transparent brown. After I fire them again with gold metallic trim in an oxidation electric kiln at cone 017, the background colour turns black: the crystals change colour with a blue tone instead of the original grey. I realised this is like a strike firing. I soon discovered all sorts of colours developed in other glazes that were never possible without the strike, especially ones with any form of titanium. This is my secret, which I no longer need to keep.

I was once supplying ten galleries, producing like a small factory. It is finally getting to me. My wrists are threatening me with carpal tunnel syndrome, and the clay company in Alberta that made some of my clays can no longer make my favourite, "porcelaneous stoneware". Certain materials used to produce this clay are depleted. Also, over the last several years, prices for raw materials have skyrocketed.

After 37 years of making pottery, I'm looking forward to doing other things I've always wanted to do but never had the time. And part of that may be just doing nothing! Making pottery has been a wonderful life and I have been incredibly lucky.'

BELOW Urn © Bill Boyd.

BELOW RIGHT Vase © Bill Boyd.

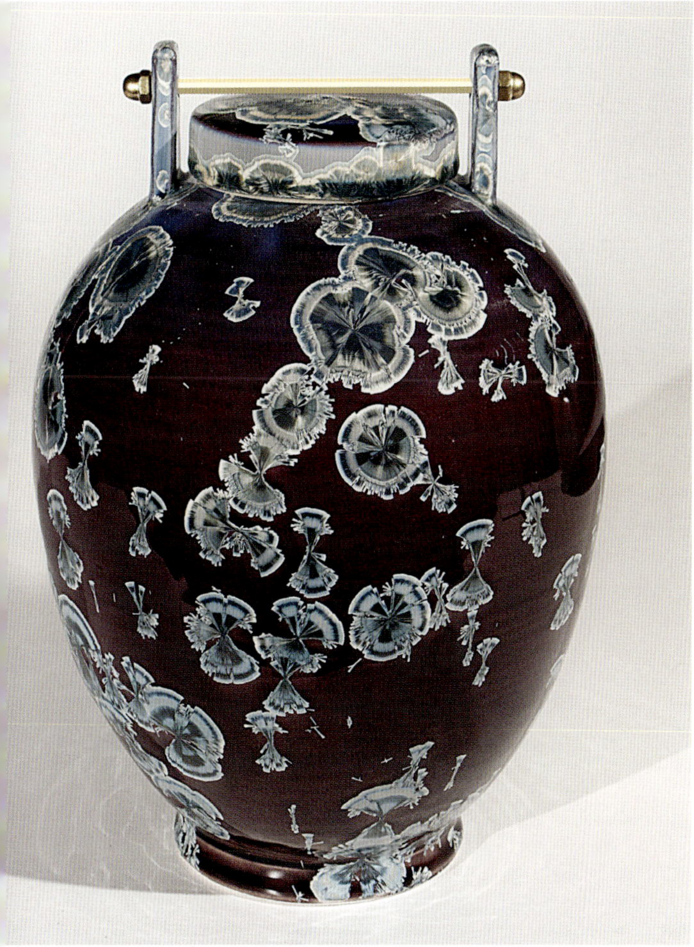

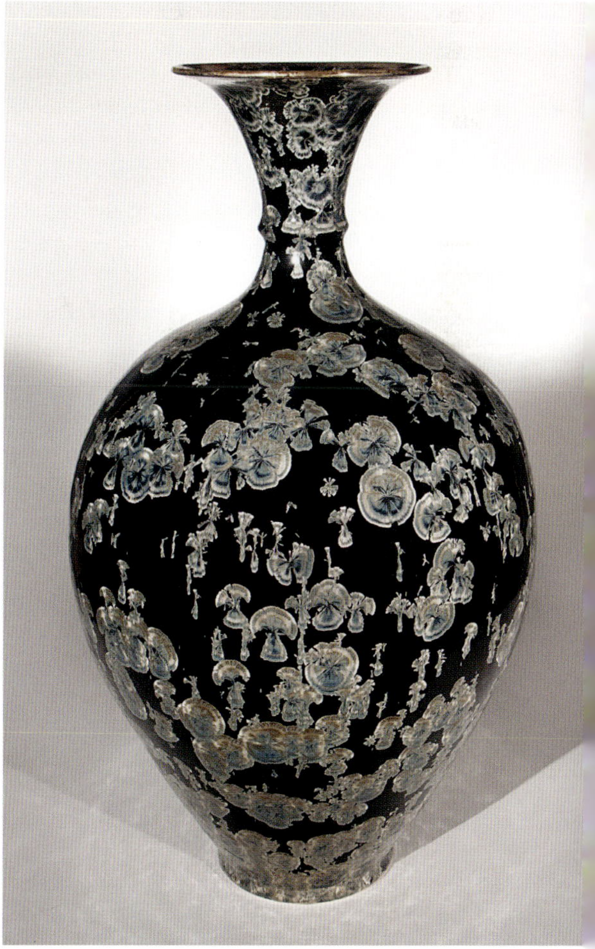

Annie Cap, England

Alongside studying international business, Annie Cap earned a fine arts degree while living in Washington State, USA. Her career in telecommunications led her to London 26 years ago. She now has dual citizenship and resides in the beautiful countryside near Canterbury.

In 2010, her English husband surprised her with a used kiln and potter's wheel. While searching for an online ceramics class, she discovered a potter holding a spectacular crystalline platter. Her career in crystalline glazes began.

The tipping point came when Annie attended a five-day workshop at José María Mariscal's studio in Spain, which focused primarily on gas reduction of crystalline glazes. She explains:

> *'To achieve bold contrasting colours, I learned I would need a reduction atmosphere. But I only had an electric kiln! Even though I was concerned about safety and the impact reduction would have on my kiln, I had to try it.'*

Using other people's alcohol-drip methods, she felt that the results were dull and unappealing. She wanted the dark colours José Mariscal had achieved. After testing various alcohols and IV lines, Annie found that 99.9 per cent pure methanol gave better, more reliable results. Methanol is introduced anywhere from 1111°C down to 1046°C (2032°F–1915°F) during the later stages of crystal growth. Methanol burns cleanly, as it contains only one carbon molecule and requires less oxygen to burn. Annie says that she has never had any carbon discoloration inside the kiln or on the pottery when using methanol.

Although she reduces at similar temperatures to those using means other than methanol, she discovered she can reduce for less time. Her range is from 17–55 minutes, with 22 minutes being her preferred duration. She doesn't use an oxygen sensor; the flame exiting the kiln is her indicator. 'I use this same technique for low-temperature crystalline lustres and for some of my copper reds', Annie says. 'In these cases, the flame is not visible, and I reduce for ten minutes.'

Annie works with a porcelain-smooth stoneware called Millenium White. 'Twilight', one of her favourite glazes, can give streaked or mottled blue to solid black backgrounds, with black, white, silver, blue, and taupe crystals, depending on the reduction temperatures and clay body.

Twilight Glaze: Cone 10

INGREDIENT	PARTS
Frit 90208 (Austrian frit. Can substitute Fusion Frit F-644)	21.15
Frit 3110	21.15
Keramikos frit 15.10 (Dutch frit. Spanish equivalent is frit VTR109)	5.67
Zinc oxide	24.17
Silica	18.13
Bentonite	1.95
Rutile	5.13
Ilmenite	2.65
Total	100.00

Firing schedule (electric kiln oxidation and reduction)

RAMPS	TEMP °C	TEMP °F	HOLD*
150°C (302°F)/hr	1144	2091	0
60°C (140°F)/hr	1250	2282	30 min or cone 10
Ramp down	1099	2010	1.5 hr
Ramp up	1117	2043	11 min
Ramp down	1022	1872	17 min
Ramp up	1097	2006	10 min
Begin reduction.			
Ramp down	1046	915	Reduce for 22 min

*Hold times indicated are averages.

'Dramatic juxtaposed colours and patterns are what I hope distinguishes my work,' Annie explains. 'After 14 years of crystalline glaze development, my ceramics finally represent me and my life. Now it's time to teach my method to others.'

RIGHT Round forms. First fired in oxidation then reduced at higher temperature © Annie Cap.

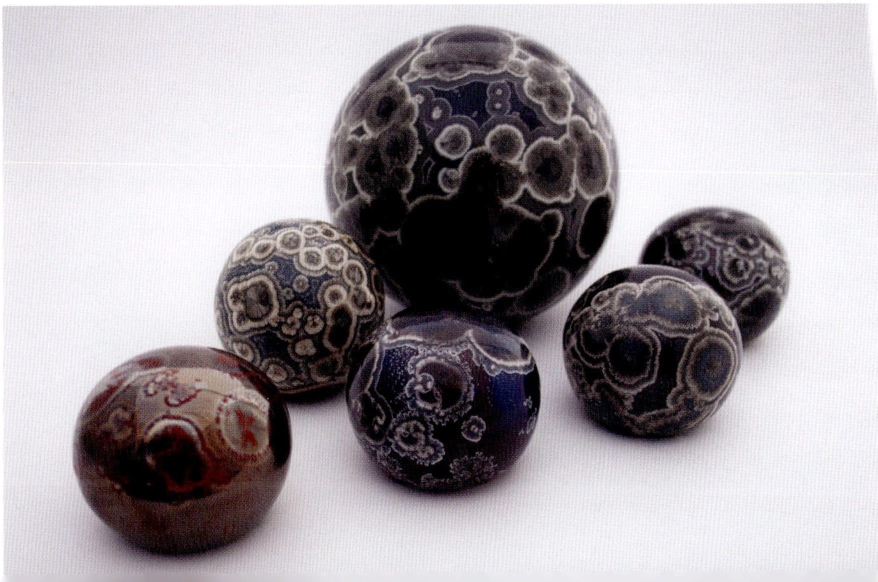

Ian Childers, USA

Ian Childers, a professor of ceramics at Mississippi University for Women in Columbus, Mississippi since 2011, has been working with crystalline glazes since 2001. He makes most of his pottery at the university:

> *'Part of my teaching philosophy is to lead by example, so I run my personal work alongside my students. I do have my own electric L&L, JH Series Crystalline kiln, and I also have use of two other university kilns that I share with the students.'*

Although Ian uses the commercially available Coleman Porcelain, he also makes his own clay that he uses for both casting and throwing. He almost perfected the formula when some of the ingredients disappeared from production. This necessitated reformulating the recipe, a time-consuming process entailing much testing. He doesn't want to share the recipe until he gets it just right.

Ian's favourite glaze recipe is as follows.

Ian Childers Base: Cone 10

INGREDIENT	PARTS
Frit 3110	48.00
Zinc oxide	20.00
Silica	23.00
Talc	5.00
Whiting	4.00
Total	100.00
Add:	
Copper carbonate	3%
Cobalt carbonate	1%
Manganese dioxide	1%

Firing schedule (electric kiln oxidation)

RAMPS	TEMP °C	TEMP °F	HOLD
Ramp up	1293	2360	0
Ramp down	1150	2100	2 hr
Ramp down	1090	2000	1 hr
Ramp down	1010	1850	1 hr
Ramp down	960	1175	30 min

If Ian is doing a post-fire reduction, he doesn't start reduction until after the initial firing is completed, which is during the final crystal growth hold. He begins reducing on the ramps down at around 1038°C (1900°F) and then continues to reduce until 538°C (1000°F).

About ten years ago, Ian bought a Fallonator, an instrument invented by another crystalline glaze potter, Terry Fallon, which is no longer in production. Essentially, it's a reduction kit that is bolted onto an electric kiln. A small burner attached to a porcelain tube goes into the bottom peephole and introduces propane into the kiln. An oxygen probe inserted in the top peephole leads to a small computer which regulates the atmosphere by increasing or decreasing the amount of propane.

Although Ian shows his work at several galleries that represent him, most of his marketing happens over social media. 'I've stuck with crystalline glazes because of the challenge. Pursuing quality and control has kept me excited for more than two decades.'

BELOW Both © Ian Childers.

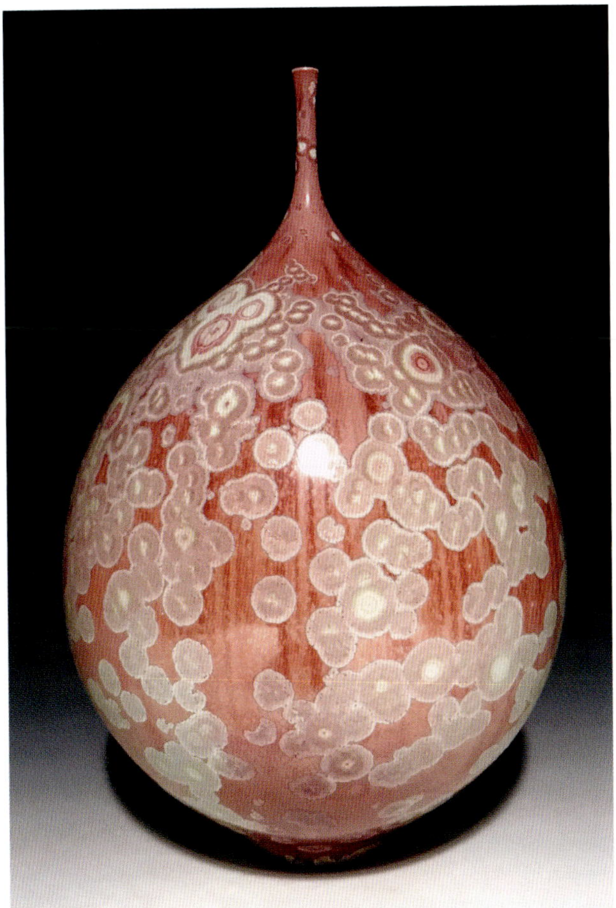

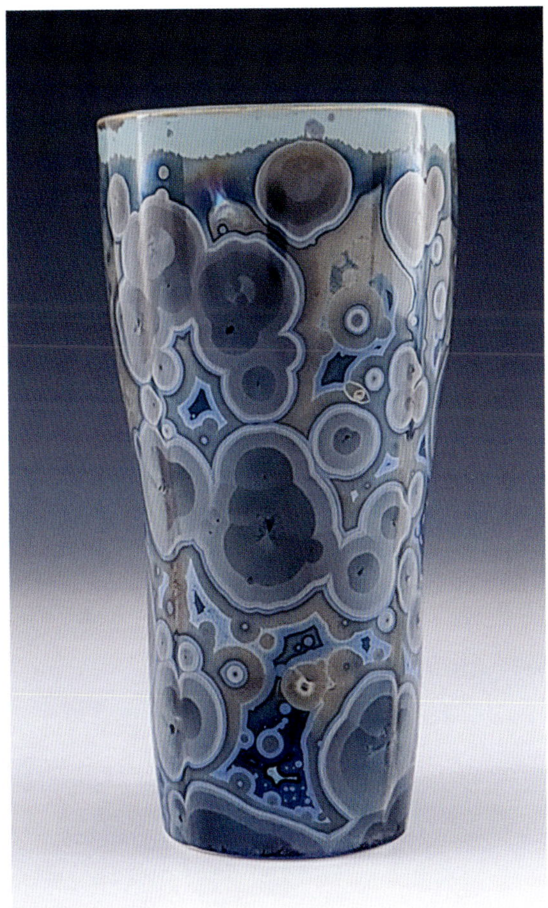

Pat Geenty, England

Pat Geenty was trained as a potter by the Benedictine Monks of Prinknash Abbey, Gloucestershire. He taught ceramics for several years at a secondary school, but because of growing family commitments changed his career to policing where he was promoted to Chief Constable of Wiltshire. During his time with the police, he also had a workshop in the garage of his home where he made functional stoneware using traditional glazes.

When Pat retired from policing, he set up a small studio in his garden in the village of Farthingstone, Northamptonshire. His interest in crystalline glazes was sparked after seeing the work of another crystalline potter, Matt Horne, and he has spent the last several years trying to perfect the technique. 'It has become an all-consuming passion for me because of the infinite variety of crystals that can be created on the surface of these amazing glazes.'

Pat mainly makes smooth, rounded forms – vases, urns, bottles, and wide, low bowls – that display crystals at their best. He purchases his clay, PF570 White Stoneware, from Valentine Clays; it is also sold by other UK ceramic suppliers. Grog and molochite are added during the manufacture which give the clay tooth for when he throws tall or wide-bodied shapes.

To prevent overnucleation of the crystals and ensure a glass-like finish, he trims and then burnishes the surface of his leather-hard pots. When the pots are dry, he sprays two coats of white porcelain slip on them, before bisque-firing to 1000°C (1832°F). This further reduces nucleation sites. 'Crystalline glazes perform best on porcelain, which is why I spray them.'

Pat makes a one-piece catch bowl for each pot. China clay is mixed 50/50 with alumina and a little wood glue, and painted onto the top of the catcher rim. This usually results in the pot coming off the catcher with a slight tap. 'I use a simple base recipe which I brush onto the surface of the bisque-fired pot,' Pat says, 'thicker at the top and middle, thinner near the base.'

Pat's Base Glaze: Cone 10

INGREDIENT	PARTS
Frit 3110	518.00
Zinc oxide	245.00
Silica	144.00
China clay	12.00
Titanium dioxide	50.00
Lithium carbonate	5.00
CMC (organic gum)	5.00
Total	979.00

'I experiment all the time, adding different oxides in varying amounts to create new colour pallets and crystal formations', Pat explains. 'Sometimes I replace the titanium with light rutile. If I want a transparent glaze, I leave the titanium out altogether.'

Firing schedule

RAMPS	TEMP °C	TEMP °F	HOLD
200°C (360°F)/hr	1000	1832	10 min
Ramp up	1260	2300	15 min
Ramp down	1080	1976	30 min
Ramp down	1060	1940	2 hr
Ramp up	1080	1976	30 min
Ramp down	1050	1922	1 hr
Ramp down	1010	1850	1 hr

'I don't do reduction firings. I use an electric Skutt KM-1027 kiln with a light load for my glaze firings. The computerised controller means that I don't have to babysit the kiln, although I always check visually at regular intervals. Most of my work is sold online via my website. I also attend a number of craft shows around the UK during the summer months.'

BELOW AND RIGHT © Pat Geenty.

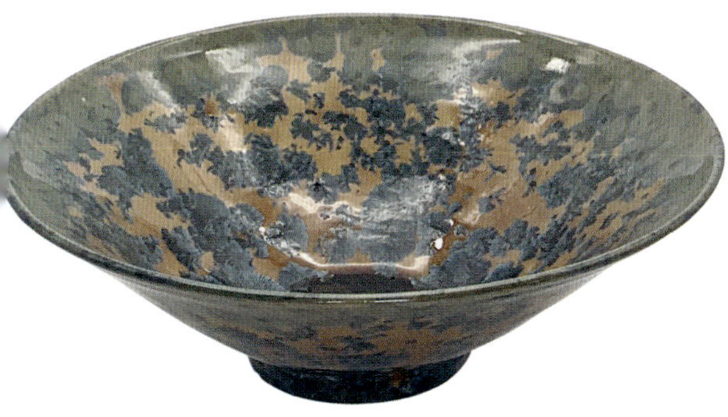

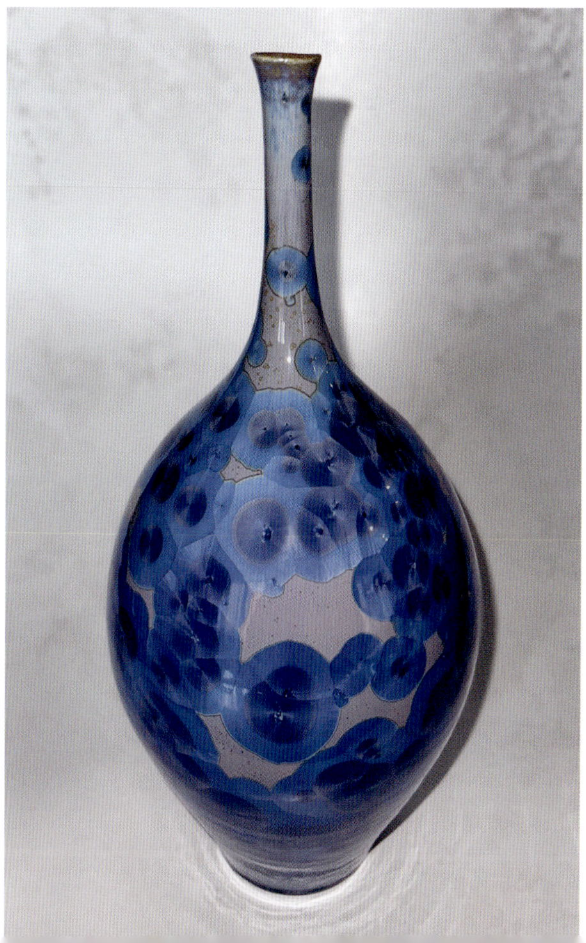

Renu Gupta, India

Renu Gupta's home and studio are in Noida, Uttar Pradesh, close to Delhi. Her studio name is Maati Maanush She has been working with crystalline glazes since 2020, after taking an online course led by José María Mariscal, as well as some offered by the Ceramic Materials Workshop. In the past, Renu used local stoneware clay, but now she works mostly in porcelain.

In India there are many vendors of pottery materials and frits, so Renu is exploring their use in crystalline glazes. However, the materials are sometimes unpredictable and might cause her clay to slump or glazes to fail. With a constrained power supply, power cuts can interrupt firing cycles. For that reason, Renu fires her crystalline glazes in a smaller-sized, Cone Art kiln made by Tucker's Pottery Supplies in Canada.

Even though there are many potters in India, Renu is the only one specialising in crystalline glazes as far as she knows. 'Most people don't understand the process, but once they see it as something different from normal glazes, they become interested and are amazed by the effort involved,' she explains. She markets her pottery through exhibitions and gallery shows, and at pottery markets in cities throughout India.

Renu's favourite glaze: Cone 10

INGREDIENT	PARTS
Frit 3110	54.37
Zinc oxide	25.41
Silica	12.53
Rutile	5.91
Black copper oxide	1.78
Total	100.00

Firing schedule

RAMPS	TEMP °C	TEMP °F	HOLD
240°C (464°F)/hr	1120	2048	10 min
100°C (212°F)/hr	1275	2327	10 min
Ramp down	1080	1976	40 min
Ramp down	1050	1922	4 hr
Ramp up	1080	1976	5 min
Ramp down	1050	1922	5 min
Ramp up	1080	1976	5 min
Ramp down	1050	1922	5 min
Ramp down	1000	1832	1 hr
Ramp down	975	1787	25 min
Ramp down	850	1562	20 min

BELOW Both © Renu Gupta.

'I enjoy the chemistry involved in making my own glazes', Renu explains, 'and love experimenting and developing new glazes both at cone 6 and cone 10 temperatures.' Renu is planning to give lessons at her studio.

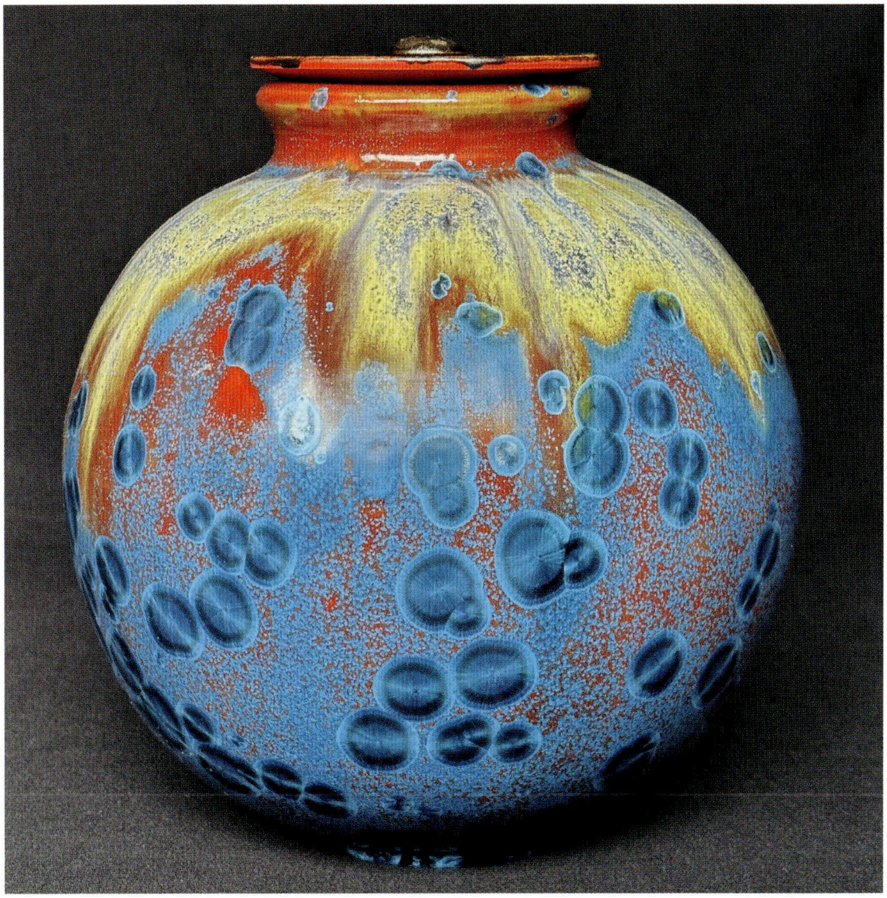

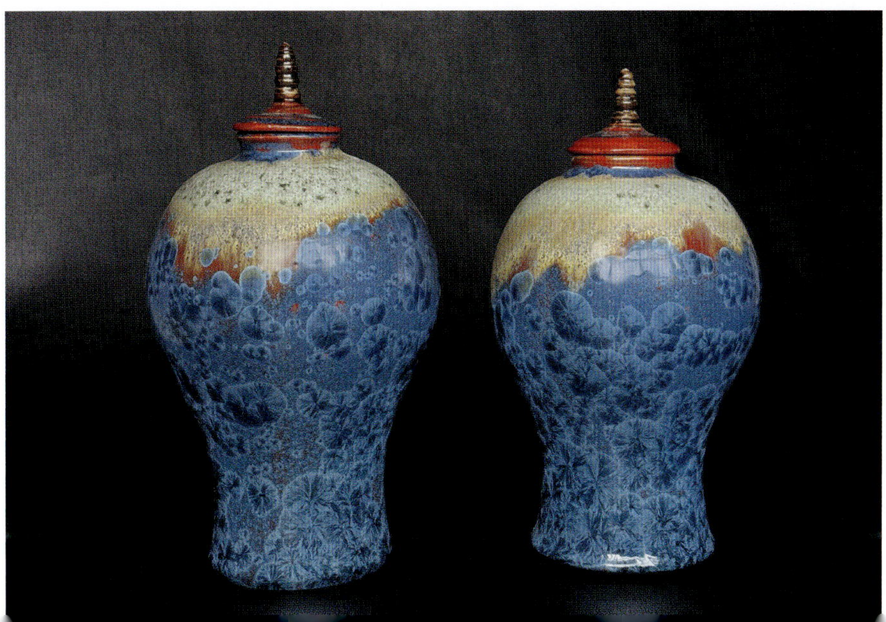

Matt Horne, England

Matt Horne has been working with crystalline glazes since 2009 at his pottery studio in West Hythe, Kent. Most of his work is thrown using Audrey Blackman porcelain. He has also used grogged porcelain for throwing larger pieces, and slip-cast porcelain for smaller items such as Christmas ornaments and mugs. Matt says:

> *'Working with crystalline glazes brings numerous complexities and challenges. Every glaze firing is both exciting and nerve-racking. I am passionate about creating unique, one-off pieces and never know what to expect. It gives me the motivation to continue experimenting with mixing my own glazes and testing different firing schedules to achieve distinctive finishes.'*

Matt's interest in these glazes was piqued after seeing an image of a crystalline vase online. Having developed many of his working practices in the age of social media, he has taken advantage of this incredible platform to share and market his work worldwide. He sells throughout the UK and ships to clients all over the world, both directly to customers and to several ceramic galleries and outlets. He also exhibits at a handful of key pottery markets, such as Ceramic Art London. Both online and off, he's found that working with crystalline glazes brings with it a community of brilliant artists with whom to share ideas. 'In my pursuit of mastering these glazes,' says Matt, 'I have drawn inspiration and influence from many other great ceramicists over the years, including John Tilton and Peter Ilsley.'

Matt's base glaze: Cone 10

This is Matt's favourite glaze. This glaze also works well in a post-fire reduction, producing metallic red crystals with a pink background.

INGREDIENT	PARTS
Frit 3110	50.50
Zinc oxide	25.80
Silica	20.40
Bentonite	3.20
CMC gum	1.00
Total:	100.90
Add:	
Titanium dioxide	6.00
Copper carbonate	2.00
Cobalt carbonate.30	0.30

Firing schedule for secondary crystals

RAMPS	TEMP °C	TEMP °F	HOLD
300°C (540°F)/hr	1180	2156	0
40°C (72°F)/hr	1265	2309	15 min
500°C (900°F)/hr	1065	1949	4 hr
500°C (900°F)/hr	1120	2048	2 hr
500°C (900°F)/hr	1000	1832	1 hr
500°C (900°F)/hr	920	1688	40 min
30°C (54°F)/hr	800	1472	30 min

'Working with crystalline glazes has allowed me to move away from production-thrown stoneware to producing unique decorative pieces, which I enjoy more as an artist. I feel lucky to be able to have worked at a job I love, and there is still so much more for me to explore with crystalline glazes over the coming years.'

RIGHT © Matt Horne.

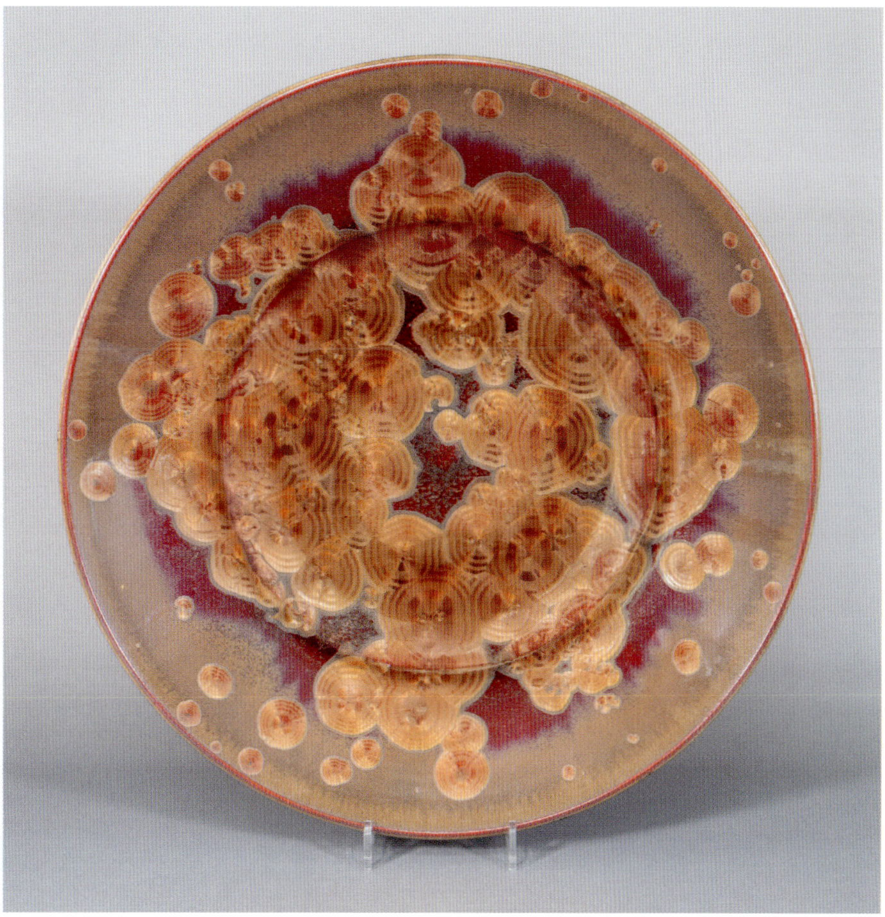

Isak Isaksson, Sweden

Isak Isaksson's home and studio are in Norrtälje, 25 kilometres north of Stockholm. Although he has been a potter for 50 years, he didn't start making crystalline glazes until 15 years into his career. He also works as a part-time family therapist. Isak was awarded Sweden's Master's certificate for his pottery skills in 2019. His characteristic crystalline glazes have become his hallmark, attracting attention in Sweden and internationally. His work has been presented in numerous exhibitions and three crystalline-glazed pots are included in the collection at the Nationalmuseum, Sweden's national gallery of fine arts.

Isak keeps a database of crystalline glazes using a glaze calculation programme to determine how much of each material to include in a specific glaze. While the programme is static, if he shifts the balance of materials in any direction, things begin to happen. Even though the programme is based on the atomic weight of ingredients, Isak has found that the actual weight of the ingredients can vary between different refractions and geographical sources, thus leading to surprises. 'I often talk online with other crystalline glaze potters. We exchange ideas and solve problems together. It's a great way to learn.'

Most of Isak's work is on porcelain. He has four electric and two gas kilns, all computerised.

Isak's favourite glaze: Cone 10

INGREDIENT	PARTS
Frit 3110	50.50
Zinc oxide	26.50
Silica	17.00
Talc	5.00
Copper carbonate	2.00
Bismuth oxide	5.00
Total	106.00

Firing schedule

RAMPS	TEMP °C	TEMP °F	HOLD
200°C (360°F)/hr	1100	2012	0
Ramp up	1285	2345	10 min
Ramp down	1055	1931	1 hr
100°C (189°F)/hr	1090	1994	2–3 hr

This glaze is often post-fired in reduction. 'When I fire in a gas kiln,' says Isak, 'I reduce as the temperature drops from 1065°C (1949°F) for one hour, or reduce at 850°C (1562°F) for one hour, depending on what effect I want.'

After a glaze firing, Isak often puts the pieces in an acid bath, using the chemical for swimming pools, pH Down. The time in the bath depends on what he's trying to achieve; glazes with cobalt require one to two hours; those with copper need up to 12 hours. The etching results in colour changes and higher contrasts.

Some of the pots are post-reduced twice. After the acid bath, they are fired in an electric kiln to 585°C (1085°F), then the gas is turned on and they are reduced to 685°C (1265°F) for about 45 minutes. Once cooled, they are again placed in the acid bath to clean off the soot. Isak says:

> *'I feel that different types of expression are suitable for different types of form. Balance is important. On a more bombastic shape, a calmer glaze feels best. If the shape is serene, then I need a bit of punch. And sometimes it's fun to just go all out and not care so much. As I see it, there is so much there, I will need about 20 to 30 more years to explore the magic of crystalline glazes.'*

RIGHT Both © Isak Isaksson.

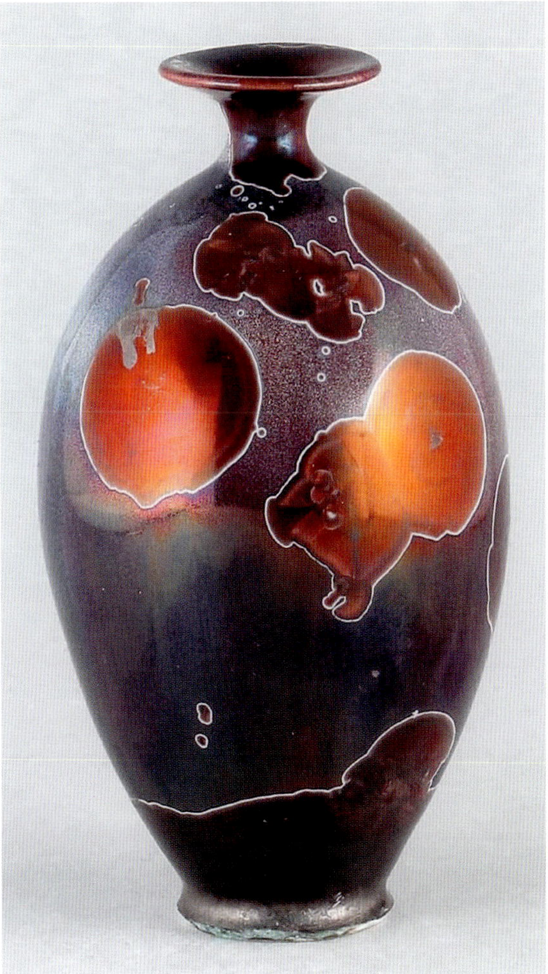

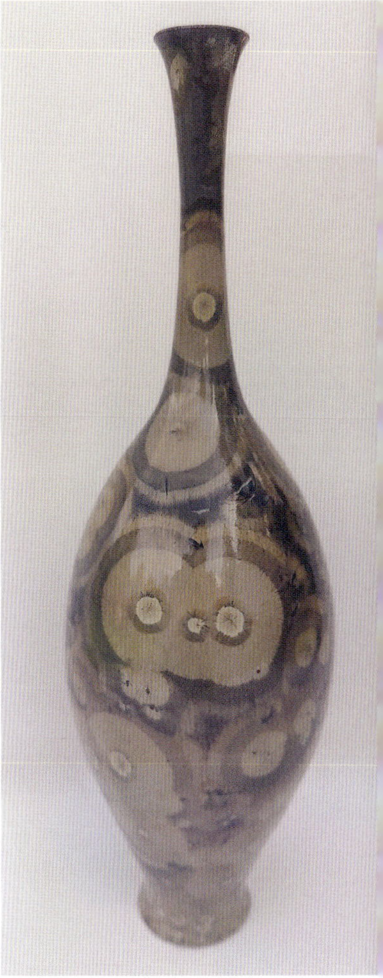

Jamie Kozlowski, USA

Jamie Kozlowski had the good fortune to be born into a pottery family. His first influences were his parents who, in 1979, built the Wizard of Clay – a large pottery studio and shop located in the town of Bristol, in upstate New York, a tourist destination in the Finger Lakes area.

His father was the first Wizard of Clay, making stoneware production items on the potter's wheel. Several employees worked as assistants while his mother ran the retail shop. Jamie explains:

> *'When I was young, I was jaded. I didn't want to work as hard as my parents even though, somewhere in the back of my mind, I knew that someday I would probably be a potter. Back then, skiing and ski instructing were my priorities.'*

In his early twenties, Jamie moved to Utah for three years to ski. Each summer, he would return home to work for his parents at the pottery, doing maintenance and other odd jobs – everything but working on the wheel. Jamie explains:

> *'One fall, I was ready to head back to Utah when I realised I needed to decide what to do next with my life. It was 1994. I thought about what might become of the Wizard of Clay when my parents retired. At that moment, I knew I had to at least see if I had the ability and patience to become a potter while my father was still active and could teach me how to throw, trim, and load kilns.'*

Having been around clay his entire life, Jamie had enough knowledge to sit down that night and make a few small pots. The next morning Jamie told his dad he was ready to get serious about making pottery.

They met each morning at eight o'clock and worked for one hour before employees arrived and the store opened. Jamie learned that he did indeed have the ability, temperament, and patience for the craft, and he improved very quickly. Eventually he exceeded his father's throwing abilities. His father credits him with being the more careful craftsman.

For several years, Jamie worked as a production potter before discovering his first crystalline-glazed pot at a craft show. Curious, he got my book, *Crystalline Glazes*, which inspired him to try this process. All the materials he needed were right there in his studio, including eight brand new computerised kilns. He started experimenting and was soon obsessed.

Fortunately, Jamie has never had to search for markets for his work. Although he has participated in a few crystalline glaze gallery exhibitions, customers flock to the Wizard of Clay's large showroom:

> *'I mostly do production stoneware, which is what the Wizard is known for. I'm not actually passionate about this kind of pottery. It's when I work with crystalline glazes that a spark ignites. I'm able to work on crystallines more during the winter,*

as our staffing and production priorities change. This time is like a holiday for me, when I can really be creative.'

Jamie has five Randall pottery wheels, all of which get used. Employees trim, attach handles, glaze, and do other tasks around the studio. His job is throwing, loading kilns, making glazes, and fixing things, be it machines, technical pottery problems, or reinforcing employee techniques. A full glaze kiln gets fired every day, while bisque kilns are fired whenever they get full enough – maybe every three days or so.

For many years, Jamie used Southern Ice porcelain from Australia for his crystalline work. Today, he uses Elaine's Crystal porcelain, available from Sheffield Pottery in Massachusetts.

Jamie has worked with many glazes over the years, and he doesn't have a favourite. The one below works well, especially with a variety of oxides. It uses a blend of three popular frits.

Jamie's glaze: Cone 10

INGREDIENT	PARTS
Frit 3110	166.66
Fusion Frit F-413	166.66
Fusion Frit F-644	166.66
Silica	250.00
Zinc oxide	200–250.00*
Spodumene	50.00
Talc	40.00
Titanium dioxide	up to 80.00
Iron oxide	20.00
Cobalt oxide	5.00

*Based on crystals to background ratio and whether titanium dioxide or high fluxing oxides, such as copper carbonate or manganese are used.

This glaze also works well over a porcelain englobe containing up to 10 per cent encapsulated stain. Jamie now has 11 Skutt kilns. Four are used for crystalline firings: a small test kiln, and three larger ones. They all have 3-inch brick interiors, APM elements and Type S thermocouples. If he wants to do post-reduction, he uses an electric kiln with an alcohol drip.

Firing programme

RAMPS	TEMP °C	TEMP °F	HOLD
Ramp up	1178	2151	0
60°C (108°F)/hr	1290	2355	0 (cone 10 down)
165°C (297°F)/hr	1107	2025	0
1.6°C (3°F)/hr	1087	1989	0
Ramp down	1037	1899	5 hr 4 min

This programme grows nice halos. The 60°C (108°F) rate climb establishes a consistent cone bend and can then be used to accurately calibrate the Type S thermocouple to a more accurate temperature setting, which can be offset to read a more accurate temperature on the control panel.

'The Wizard of Clay is a very busy place', Jamie says. 'That's why I enjoy the quiet of winter when our production line changes. This allows me time to squeeze in as many crystalline firings as possible.'

LEFT AND BELOW
Both © Jamie Kozlowski.

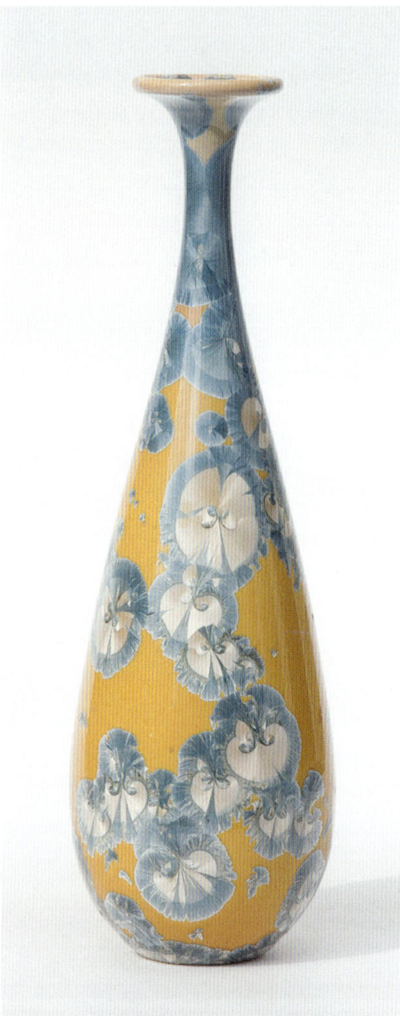

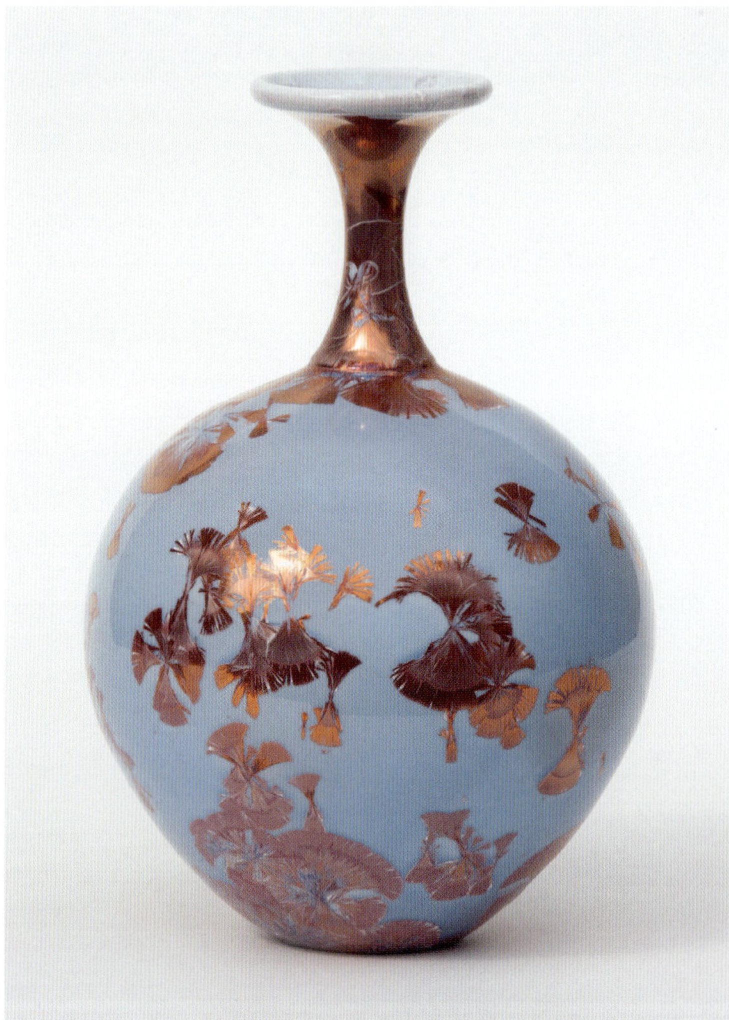

Andrew Krueger, USA

Andrew Krueger works in the basement studio of his home in Sheboygan, Wisconsin. Initially, his work focused on cone 6 production ware, but he transitioned to cone 10 decorative crystalline during the COVID-19 pandemic. He discovered that the new focus satisfied his need for both artistic expression and scientific exploration. All of Andrew's work is made on the potter's wheel using a blend of Midwest porcelains. He sells his work through galleries and fine-art fairs across the US. Andrew says:

> *'Since most of my glazes are quite bright and striking, I tend to lean towards simple, symmetric shapes with clean, flowing lines. A complex, eye-catching glaze can feel a bit overwhelming on a busy form. In a quest to find the best possible surfaces for my pots, my palette of glazes is constantly evolving, and I am continually testing new glazes and refining current ones.'*

Andrew has been testing glazes over numerous refractory stains to achieve unusual colours in the background, accentuated by dynamic crystals. All his current tests are focused on understanding both how glazes interact with stains and how they can work to achieve bright colours, while still maintaining a refined, elegant look.

Andrew sprays the stains over dry greenware. He wants the colour of the stain to show through, so he uses a glaze that minimally opacifies. He often uses small amounts of colourant in his glazes to alter the stained ground colour. Working with freshly mixed glazes, he prefers a brushed application and keeps batch sizes small to minimise waste.

Base glaze over stain: Cone 10

INGREDIENT	PARTS
Frit 3110	553.40
Zinc oxide	220.00
Silica	184.10
Talc	31.40
Titanium dioxide	16.00
Peptapon 9 (suspension stabiliser)	7.00
Total	1011.90

Adding colourants also works well.

Results over red stain

Ingredient	Proportion	Result
None		Bold red ground with white/slightly pink crystals
Copper carbonate	2%	Maroon and purple ground, with silvery mint-coloured crystals
Manganese dioxide	2%	Bright red and pink ground with rose gold crystals
Cobalt carbonate	0.5%	Purple ground with blue crystals

Refractory stains significantly reduce the crystal growth rate, so the firing schedule for these glazes is quite long. Andrew fires in oxidation in an L&L top-loading kiln.

Firing schedule for crystalline recipe over stain

RAMPS	TEMP °C	TEMP °F	HOLD
38°C (100°F)/hr	121	250	5 min
204°C (400°F)/hr	1149	2100	0
42°C (108°F)/hr	1268	2330	13 min
121°C (250°F)	1052	1925	2.5 hr
260°C (500°F)/hr	1099	2010	6 hr
7°C (45°F)/hr	1038	1900	1 hr
7°C (45°F)/hr	1017	1862	45 min

'The magic of crystalline glaze is seeing how far it can be pushed and manipulated in different directions. I still learn something new about these glazes every time I fire my kiln, and I know that, in 40 years, I will likely be able to say the same thing. There are just so many possibilities.'

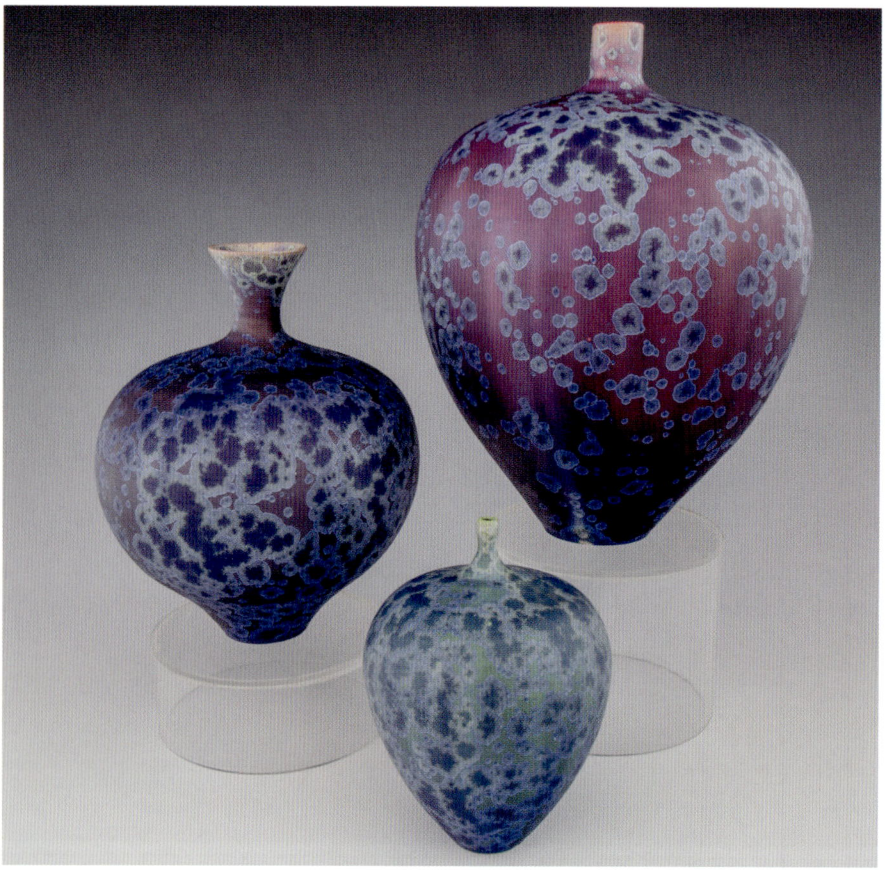

LEFT © Andrew Krueger.

Kate Malone, England

Kate is one of England's foremost and most popular ceramicists. Whether it's a small, affordable piece, a large, sculptural decorative form, collaboration with architects, or a seven-storey, award-winning building façade covered with crystalline tiles in London's Savile Row, her characteristic style is exceptional.

Kate has three distinct ways of working with clay: studio pottery production work, large-scale public projects, and glaze research. Nature is her influence; from fruits, vegetables, nuts, and seeds to various organic shapes, all are highly decorated in vibrant, colourful crystalline glazes. She explains:

> *'Crystals are part of my intent, but not the major effect. They reinforce my objective: to portray nature and the "life force" as inspiration. It's fascinating the way crystals growing within a glaze imitate geological activity. Glazing is alchemy and growing crystals seems like an extra bonus. This extraordinary process adds to the fact that my subject matter aims to capture some part of the magic and miracle of nature. It's a reaffirmation that an audience hopefully notices once the form has arrested their attention. It's quite difficult – given the separate processes of making, bisquing, glazing and firing – to ensure that there is a true balance between form and surface.'*

The majority of Kate's work is now produced in her home studio in Sarre, Kent, although London was her primary studio and home for 30 years. This space continues to display the crystalline glaze archive, and hosts meetings with architects, collectors, interior designers, fellow charity co-founders, students, and interested visitors.

Kate has two full-time assistants and many students and part-time helpers, depending on their and her needs at the time.

Glaze chemistry fascinates her, and through research, Kate has accumulated over 2,000 crystalline glaze recipe variations, which she has collected into a manual for herself, students, and other interested parties.

When commissioned to decorate a seven-storey building, Kate invited ceramicist Rich Miller to join her. Together, they produced over 11,000 ceramic tiles, all with crystalline glazes in either white or black. The bisque-fired tiles were made in Germany from an extruded clay, which Kate and Rich tested for two years to perfect. Each was brush glazed and set into the kiln with a spirit level because of the fluidity of the glazes. This commission won the project first place in the World Architecture News (WAN) Awards.

'I am fascinated by the way a crystalline glaze refracts light and appears to be constantly changing,' Kate says. 'Each tile has a dense blanket of crystals on the surface.' Kate and Rich worked for four years on this project. Later, Rich succeeded Kate as a judge on the TV show, *The Great Pottery Throw Down.*

In 2019, Kate was appointed a Member of the Order of the British Empire (MBE) for her service to ceramic art. For the last 30 years, she has sold her work exclusively through the art dealer Adrian Sassoon, whose gallery shows her work in Europe, the US, and the UK.

Kate uses a clay called T Material, made by Potsclay. It's highly refractory, white-burning with low thermal expansion, and contains several different grades of molochite and granular kaolin, making it ideal for large sculptures and hand-building.

All of Kate's glazes are essentially crystalline. She likes to use different frits; some render many crystals and others very few. Crystal production might increase if she were to cool the kilns more slowly, but she chooses not to be that specific.

Her favourite glaze recipe is a 1.5 per cent black nickel oxide addition that results in beautiful crystals with varying effects. Some are electric blue in a rich, honey-treacle transparent base. 'Even though I have over 2,000 recipes, this one remains the most interesting and is always mixed and ready to use.'

AF2 Glaze: Cone 10

INGREDIENT	PARTS
High alkaline frit	58.00
Zinc oxide	23.00
Silica	21.30
Bentonite	2.00
Black nickel oxide	1.50
Total	105.80

Firing schedule

RAMPS	TEMP °C	TEMP °F	HOLD
61°C (110°F)/hr	181	358	0
81°C (146°F)/hr	301	574	0
101°C (182°F)/hr	601	1114	0
121°C (116°F)/hr	1263	2305	1–5 min

Kate uses a Stafford controller to fire her three electric kilns. She fires them slowly to maximum temperature because of the large size of some of her pieces. By packing her kilns as tightly as possible, she hopes to reduce their environmental impact, while the density of the kiln allows for natural slow cooling, enough to grow the crystals.

Without a specific cooling cycle, Kate still gets dramatic crystal growth. A lot of her forms are enclosed shapes, and the heat that has taken several hours to build up inside them cannot quickly escape. The pieces themselves provide a slow cooling. Kate explains:

> *'The fact that crystal glazes are very fluid and run over and around contours is something that I use to help emphasise the forms that I create. The crystals are not the primary essence of the whole piece, but one aspect. My objective is to balance form and surface.'*

BOTH Image courtesy of Adrian Sassoon Gallery. Photography by Sylvain Deleu © Kate Malone.

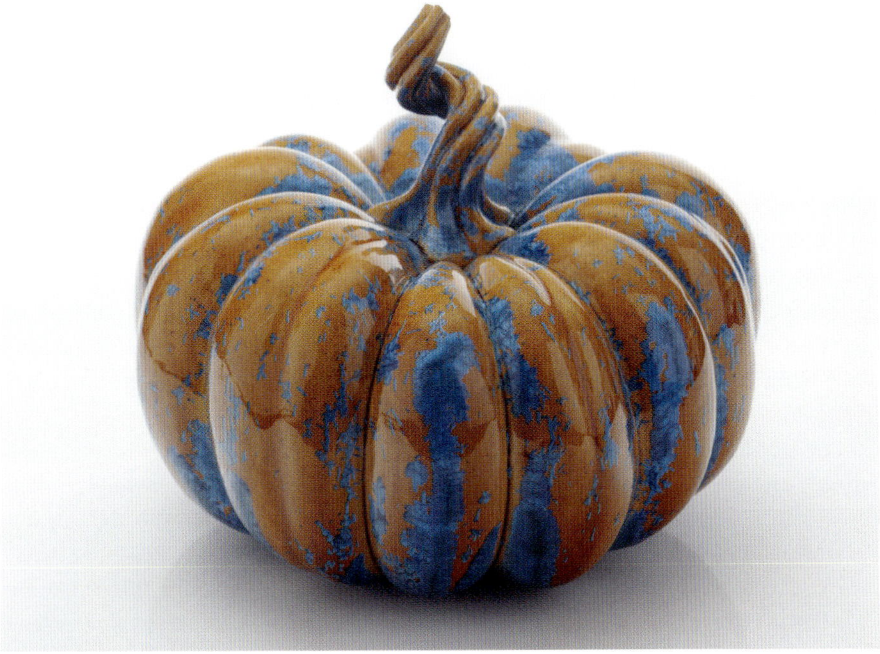

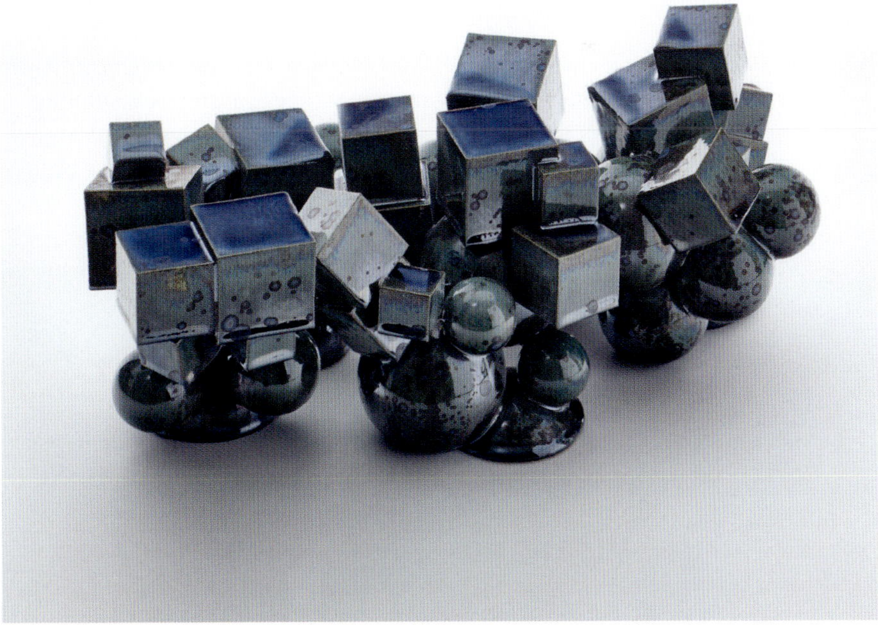

José María Mariscal, Spain

José Mariscal was born in La Bisbal d'Empordà, an important ceramics town in Catalonia. Pottery has been the main economic activity of the town since the eighteenth century, although pottery making has taken place there since 1502. José learned to throw from his father, a potter who spent his whole career throwing for others.

In 2003, José and his wife, Maite, rented a studio in La Bisbal, where he made functional, low-temperature pottery. Back then, he fired his work in a diesel-fuelled, 2 m (6½ ft) square kiln. He also did some raku and terra sigillata.

Five years later, he and Maite bought the house where they now live and work, which has a large studio on the first floor. Maite takes care of clients, organises markets and exhibitions, and manages his publicity, which includes outreach on social media such as Instagram and Facebook.

José first saw a potter at a market selling crystalline glazes in 2010. 'I was instantly hooked and researched the subject,' he explains. 'The never-ending possibilities that these glazes offer has inspired me to continue in this media. I've gained a great deal of knowledge through websites and correspondence with other crystalline glaze potters.'

The International Congress of Zinc Crystalline Glazes, organised by José, took place in La Bisbal in 2013. During the congress, Ian Childers and Phil Hamling demonstrated how applying an acid on the finished piece eats away part of the glaze surface, producing a spectacular effect. José says:

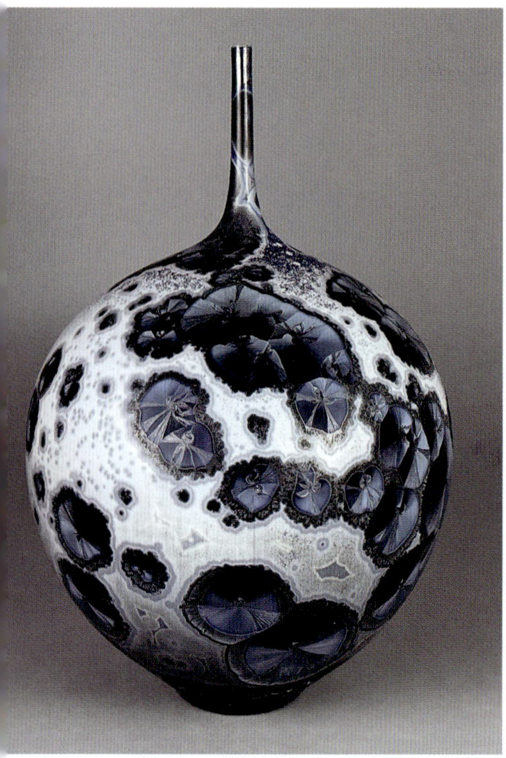

> *'After much research, I developed a glaze and achieved the results I was looking for. Although I call the glaze Galaxy, it has many variations. All of them have in common that they are copper saturations and have an acid bath after the firing.'*

José uses sodium bisulphate as his acid and keeps the pot in the acid bath for about 12 hours.

LEFT María Mariscal
© Jose MARISCAL Paneque.

Galaxy Glaze: Cone 11–12

INGREDIENT	PARTS
Fusion Frit 644	460.00
Zinc oxide	265.00
Silica	270.00
Bentonite	10.00
Titanium dioxide	10.00
Total:	1015.00
Add:	
Red copper oxide	45.00
Manganese dioxide	20.00

Galaxy Glaze Variation: Cone 11/12

INGREDIENT	PARTS
Frit 1510 or VTR 109 (If using frit VTR 109, use a little less silica)	510.00
Zinc oxide	245.00
Silica	250.00
CMC gum	10.00
Titanium dioxide	10.00
Total:	1025.00
Add:	
Copper oxide	45.00
Copper carbonate	45.00
Cobalt carbonate	8.00

Firing schedule

RAMPS	TEMP °C	TEMP °F	HOLD
6 hr climb	1282	2340	20 min
Ramp down	1085	1985	3.5 hr
Ramp down	1045	1913	1 hr
Ramp down	1000	1832	1 hr

When the pots come out of the kiln, they are metallicized black due to the copper saturation and must be immersed in acid. The more time in acid, the cleaner the background. By modifying the firing programme, he gets more halos and secondary crystals.

During the COVID-19 pandemic, José started teaching online. He now gives workshops and teaches throwing and crystalline glazing across Europe, North America and Australia. He also sells to markets in Europe, both at exhibitions and through his website. 'My interest in glazing has led me to try different types, such as lustres and Shino glazes', José says. 'I always want to learn more about these fascinating glazes.'

Holly McKeen, Canada

Holly McKeen's studio is located on her farm in Chilliwack, British Columbia. Twenty-five years after she attended art school, she was finally able to become a full-time potter. While creating a line of functional stoneware, she honed her throwing and design skills.

Holly is semi-retired from her active pottery business and on-site gallery, and now has more time to explore the alchemy of crystalline glazes, as well as strike firing, acid etching, and the layering of multiple glazes. In the last few years, she has also experimented with adding molybdenum oxide to crystalline glazes for iridescent effects, and lustre variations. She has also worked with aventurine crystals, high-manganese glazes, and post-fire reduction lustre reds. One of Holly's specialities is strike firing. She explains:

> *'Brilliant colours can be achieved in certain glazes by following a cone 10 firing with a simple refire in oxidation to cone 017. Any base crystalline glaze with a small amount of cobalt, manganese dioxide, and over 4 per cent titanium dioxide brightens up to varying degrees following an oxidation strike, with enriched gold, orange, and red colours in the background. The crystals also become more alive and animated following the strike. A fairly thick application of glaze is recommended for optimum results.'*

Not all oxide additions result in a colour burst. However, when it does happen, it results in brighter crystals and an enhanced background colour that can be quite dramatic.

Holly has also created impressive 'Nebula' glazes with colourful red backgrounds, using silver nitrate followed by an additional oxidation strike fire.

The exploration of complex, high-fired crystalline glazes enables her to offer unique work for her customers, who can view it at her home studio or online. She now gives workshops on request where she shares her glazes and years of experience.

McKeen's 413/644: Cone 10

INGREDIENT	PARTS
Fusion Frit F-413	58.00
Fusion Frit F-644	8.00
Silica	16.00
Zinc oxide	24.00
Talc	2.00
Spodumene	2.00
Veegum T	0.60
Total	110.6
Add:	
Cobalt carbonate	0.7%
Manganese dioxide	3%
Titanium dioxide	4.5%

Holly uses many different base glazes for strike firing, but this is her favourite. 'To my eye, it's not only about the crystals', Holly says. 'I especially enjoy the blending of glazes with colourful background effects that trigger the imagination and create landscape images open to individual interpretation.'

LEFT © Holly McKeen.

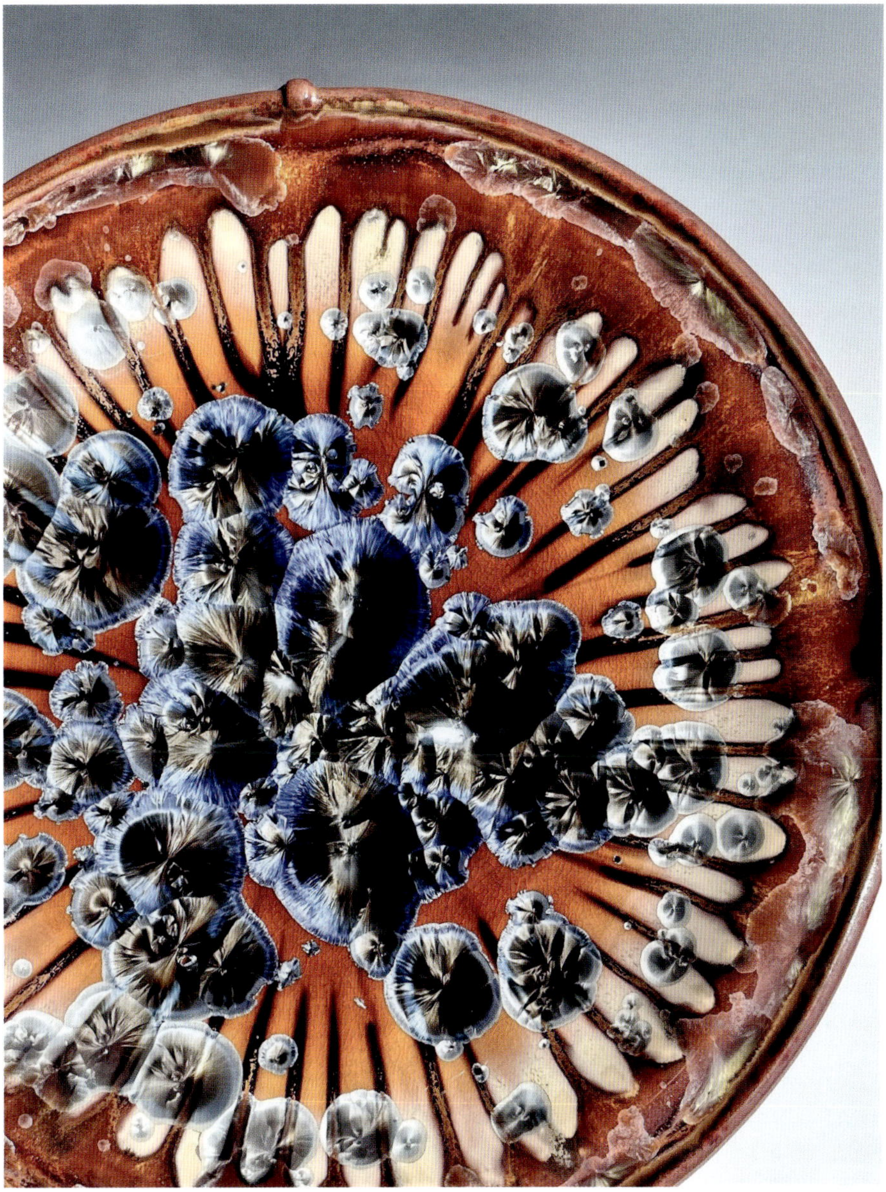

Bill Powell, Australia

Bill Powell's studio sits on an acre of land at the base of the Great Dividing Range, west of the Gold Coast in Queensland. It is equipped with 12 wheels: one for himself, the rest for teaching classes and workshops. He also has five kilns, both gas and electric. He sells his ceramic art under the title of Bill Powell Art & Design, while teaching and giving workshops under the name CeramX School of Clay & Glass. He explains:

> *'I began exploring crystalline glazes in 1977 after reading Herbert Sanders' book,* Glazes for Special Effects, *where one chapter is dedicated to crystallines. I have pursued these glazes ever since. I find the ever-varied nature of these seductive surfaces fascinating and was immediately hooked.'*

Bill uses a range of white, cone 10, commercial stoneware clays, and his porcelain choice is currently an Australian porcelain body called Lumina from Keane Ceramics. It's used for both throwing and coloured slip decoration.

He also does post-reduction of crystalline glazes. Recently he built a small, single-burner, fibre gas kiln for post-reduction of crystals and other high- and low-temperature firing techniques.

> *'One of my favourite glaze recipes I call "Moss Green". This glaze exhibits strong, green crystal formations with variegated black outer halos and a collection of secondary crystals in the background. It's reliable and it often has multiple formation habits depending on temperature and time.'*

Moss Green Glaze: Cone 10

INGREDIENT	PARTS
Frit 3110	480.00
Silica	250.00
Zinc oxide	220.00
Bentonite	10.00
Titanium dioxide	5%
Red copper oxide	5%

Firing cycle

Ramps	Temp °C	Temp °F	Hold
Full power	1285	2345	
Ramp down	1080	1976	3 hr
Ramp down	1040	1904	2 hr

Bill has been represented by Redhill Gallery, Brisbane for the past 12 years. 'I take great joy', Bill says, 'in exploring the endless potential of forms, firings and formulae of both gloss and matt crystalline glazes and sharing these with others.'

RIGHT Both © Bill Powell.

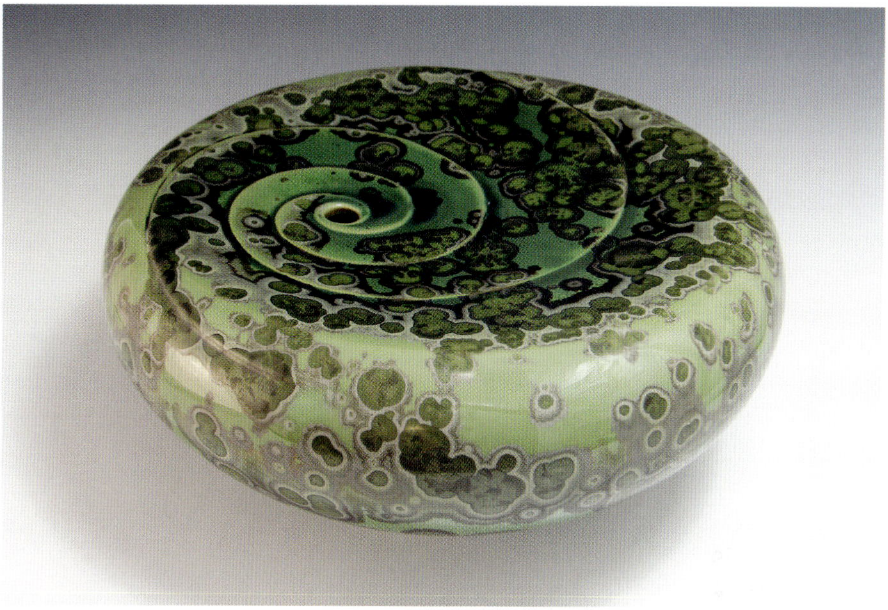

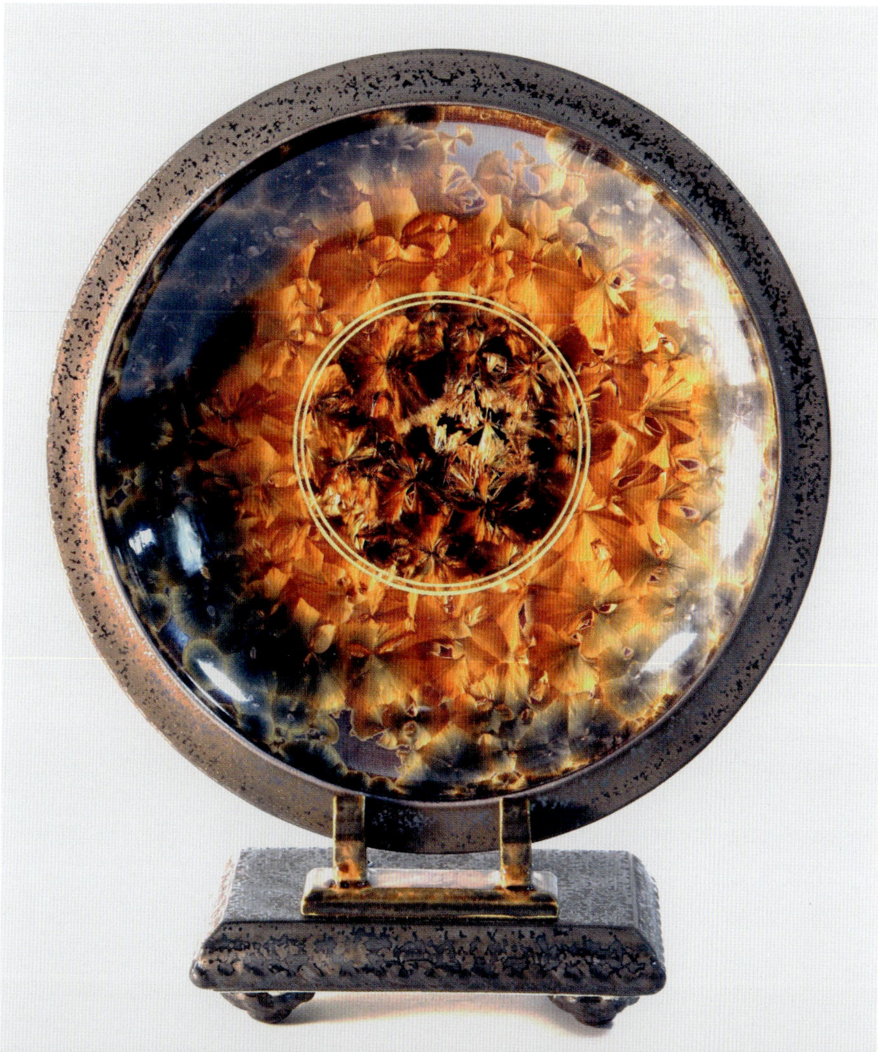

Ted Secombe, Australia

Ted Secombe lives in the Yarra Valley, an area of outstanding natural beauty and a major wine-producing region about one hour from central Melbourne, Victoria.

His studio is modest in size, about 50 m^2 (538 sq ft). Beyond his throwing room is an expansive verdant valley, giving way to a eucalypt-covered mountain range. Late afternoon skies present vibrant and pastel-coloured vistas. The magnificent view provides an ever-changing inspiration for glaze ideas.

Ted uses porcelains exclusively, with a porcelain from Keane Ceramics being the clay body favoured by most Australian potters. On occasion, he does import some Japanese porcelains. Ted says:

> *'In past years, I've embraced the commercial world, producing big volumes and monumental works for 5-star hotels, casinos, and shopping centres. The works for the commercial market were mostly crystalline glazes. The volume of work was always very substantial. For example, The Ritz Carlton in Singapore purchased 640 pieces for the hotel: 600 slip-cast porcelain pieces with satin-matt crystalline glazes – one for each room – and 40 monumental works for public areas of the hotel. The largest works commissioned for Asia were in the vicinity of 170 cm (5 ft 7 in) high. I have worked with clients as far away as the Middle East but most particularly in Singapore, Hong Kong, and Kuala Lumpur. Pottery destined for the Asian market numbered in the thousands of pieces.'*

Ted's favourite glaze is a satin-matt recipe that he has used for over 40 years.

Matt crystal glaze: Cone 10 (electric or gas kiln)

INGREDIENT	PARTS
Barium carbonate	252.00
Zinc oxide	202.00
Potash feldspar	200.00
Silica	154.00
Soda feldspar	123.00
Titanium dioxide	60.00
Total	991.00

Small amounts of the following oxides can be added either alone or in combination, although nickel is not recommended as the results are poor. When using two or more oxides, reduce the amounts:

- Copper carbonate: up to 2%
- Cobalt carbonate: up to 1%
- Iron: up to 1%
- Manganese dioxide: up to 2%

Firing schedule

STEP	RAMP	TEMP °C	TEMP °F	HOLD
1	8 hrs to	1285	2345	10 min
2	Ramp down	1060	1940	30 min
3	Ramp up	1100	2012	15 min
4	Ramp down	1060	1940	30 min
Repeat steps 3 and 4 another 3 times.				

Ted has five kilns: three electric and two gas. He does post-fire reduction crystals in a small 0.113 m^3 (4 cu ft) gas kiln and high-temperature reduction in a 0.453 m^3 (16 cu ft) gas kiln. Ted says:

> *'I sell mostly through exhibitions, both nationally and internationally. Often, I allocate funds to promote exhibitions in my own gallery, although it is always open to visitors by appointment. Some clients prefer to purchase pieces directly, rather than going through an intermediary. This has proven to be a very successful exercise.'*

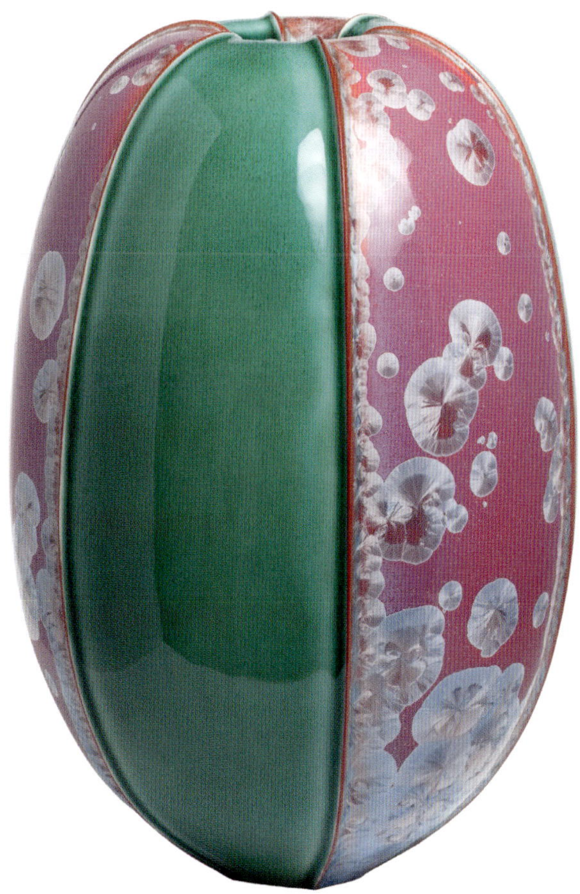

RIGHT © Ted Secombe.

Mariela Tiferes, Argentina

Mariela Tiferes's studio is in a quiet, residential neighbourhood of Buenos Aires, the capital city of Argentina. In the morning, she teaches art at a primary school; afternoons are spent offering ceramic classes to children, teenagers, and adults at her studio. Her classes have up to ten students, which allows for a comfortable working environment.

Mariela has been experimenting with crystalline glazes since 2017. She has taken several courses in Argentina and abroad. Every Saturday, she meets with her former pottery teacher, Maximiliano Abbiati.

Mariela studied at the Prilidiano Pueyrredón National School of Fine Arts (now the National University of the Arts), where she received a degree in sculpture. She continued her training with sculptor, Leo Vinci, where she worked with a variety of materials such as resin, cement, wood, and alabaster. She also attended an oil-painting workshop, which trained her to work with colour. Nowadays, she chooses clay as her medium, given the infinite possibilities of shapes, glazes, and surface treatments it presents.

Mariela developed an interest in crystalline glazes after seeing crystalline pieces by different artists on Facebook. She showed these photos to Maximiliano, an expert in the art of oriental glazes, who also teaches at the National University of the Arts. He had no experience with crystalline glazes, so they researched them together. At first, Mariela didn't create any crystals but learned a lot along the way. Then, Maximiliano introduced her to Mario Dominguez, the rector of the University of Lomas de Zamora, who gave her a recipe and a firing schedule, and she was finally able to develop her first crystals. She says:

> *'Two books were especially important for my further evolution and development in the world of crystals. They were* Crystalline Glazes *by Diane Creber and* The Art of Crystalline Glazing *by Jon and Leroy Price. During the pandemic, I also took several virtual courses with José Mariscal. I ended up sometimes having satisfactory results and other times making mistakes, having spills on the kiln shelves and other accidents.'*

However, instead of giving up, it made her want to try again. Mariela has five potter's wheels for her own and her students' use. Since Argentina doesn't have a developed pottery industry, her wheels and kiln are made on request by other potters or artisans from a range of fields.

Mariela works mainly on the wheel, and continually challenges herself with new shapes and sizes. She makes her own clay, a rose-coloured stoneware extracted from a quarry in the province of Neuquén. The pink shading is due to the iron content. That clay, called APM by the quarry, is mixed with tinkar – another clay from the same quarry, silica, and potassium feldspar.

Mariela also mixes her own porcelain. For the crystalline glazes, she uses both stoneware and porcelain. 'Some recipes work better on porcelain, while others are better on stoneware,' she states.

Her favourite glaze recipe gives her a variety of colours ranging from white to blue crystals on a dark background, with secondary crystals in brown. The blue colour comes from the iron oxide in the glaze.

Cone 9

INGREDIENT	PARTS
Frit 3110	484.00
Zinc oxide	250.00
Silica	180.00
Titanium dioxide	78.00
Alumina	10.00
CMC gum	10.00
Iron oxide	6.00
Total	1018.00

Firing schedule

RAMP	TEMP °C	TEMP °F	SOAK
Ramp up in approximately 6 hours	1263	2305	0
Ramp down	1100	2012	2 hr
Ramp down	1020	1868	3 hr

Mariela sells her work online and in local galleries and at craft fairs. Her studio also has a showroom where people can come to see her production and buy if they are interested. 'Those mornings when I am waiting to open the kiln are very intense and full of expectations', she says.

RIGHT © Mariela Tiferes.

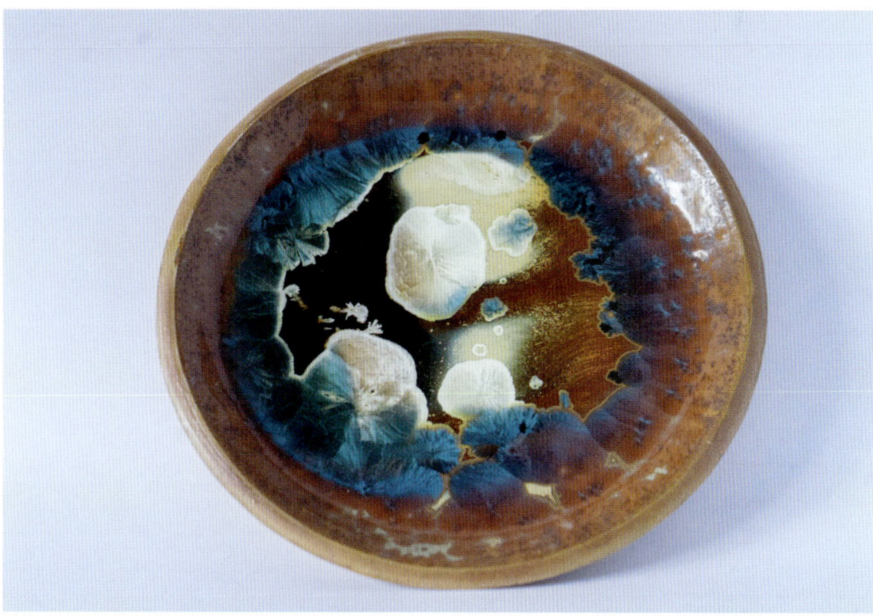

John Tilton, USA

Since 1978, John Tilton and his wife, Anne, have lived in a rural, spiritual community, ten miles from the closest grocery store, near Gainesville, Florida. John's studio is nestled in the surrounding woods. Twice each day the community meditates and chants together. He explains:

> *'Seeing how the mind becomes quiet during this time has a large influence on my life and my work. As the quiet increases, the work becomes clearer and our tiny place in the universe comes into focus. One of the main themes of my glazes is to put universes on the surfaces of my pots. I consider the shape more important than the glazes and I try to impart a spirit to the forms in hopes they communicate something about that quietness.'*

John was working on a PhD in mathematics at the University of Florida when he started making pots to relieve the intense mental strain of his studies. He decided to leave mathematics and devote his life to clay, enrolling in the University of South Florida and graduating with a Master of Fine Arts degree in ceramics in 1972.

John mixes his own porcelain clay as a slip, then screens and dries it on racks he built for the purpose. Working as a potter for more than half a century, he has seen many changes:

BELOW © John Tilton.

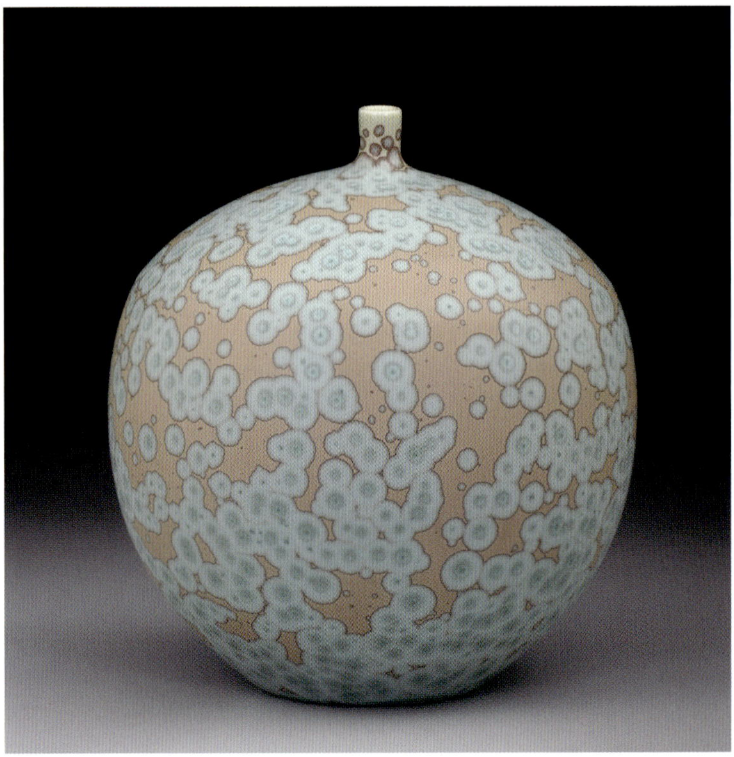

> *'For instance, our original batch of Ferro Frit 3110 has now disappeared and all the new batches I've tried – maybe 30 in all – are different from the original, and slightly different from each other. There are good batches and bad. My advice, when you buy chemicals, know the batch numbers, do the testing, and, when you get what you want, buy what you'll need for as long as you plan to work, if that's possible.'*

Marketing his work has also changed drastically, from selling at art festivals to marketing through his website with the help of Instagram and Facebook. Now John sells almost all his work through social media. The shift in sales has also changed the flow in the pottery shop, with pottery displays converted to storage areas for shipping materials. The glazes here are three of John's favourites.

Oxidation Zinc Crystal: Cone 10

This glaze needs colourants and must be fired in oxidation. Apply thinly.

INGREDIENT	PARTS
Frit 3110	134.00
Zinc oxide	100.00
Silica	64.00–82.00
EPK	12.00
Titanium dioxide	0–8%

Firing schedule

RAMPS	TEMP °C	TEMP °F	HOLD
260°C (500°F)/hr	85	185	30 min
135°C (275°F) /hr	1288	2350	Variable
Ramp down	1093	2000	40 min
Ramp down	1066	1950	30 min
Ramp down	1038	1900	20 min
Ramp down	1029	1885	10 min
Ramp down	371	700	Off

Manganese Crystal Glaze: Cone 10

Can be fired in either oxidation or reduction. This glaze has a strange coefficient of thermal expansion. The inside and outside of the pot must have the same glaze or the pot may break in cooling.

INGREDIENT	PARTS
Custer feldspar	70.00
Powdered manganese dioxide	28.00
Granular manganese dioxide	2.00
Total	100.00

John's reduction matt: Cone 10

Apply thinly. It's very runny. It also needs colourants; try copper, tin, cobalt, or pink stains. Fired together with John's stoneware pots, the reduction kiln cools very slowly but there are no holds in the temperature.

INGREDIENT	PARTS
Nepheline syenite	85.00
Barium carbonate	65.00
OM#4 ball clay	15.00
Whiting	20.00
Rutile	15.00
Total	200.00

Glenn Woods, USA

Glenn Woods first became interested in pottery while attending high school. He went on to earn a Bachelor of Science Degree in Art with a concentration in ceramics. After graduation, he held a variety of jobs. It took several years for him to gather the courage to become a full-time studio potter.

In 2001, Glenn and his partner, Keith Herbrand, bought an existing pottery business and moved from Chicago, Illinois to Palm Harbor, Florida to pursue their dream. Glenn's plan was to work full-time as a studio potter while Keith would manage all other aspects of the business. Over the years, however, Keith has transitioned into a studio potter as well.

Although they work together, Glenn focuses on decorative pieces, whereas Keith does more functional ware; they fire their pottery together using the same array of crystalline glazes. Glenn generally works on a series based on a specific theme, often carving, beading, or piercing his work to create a more interesting form. Crystalline glazes are used to accentuate these alterations. 'When I first became a full-time potter,' he explains, 'I wanted the business name to acknowledge both of us. So, we call it, "Pottery Boys Clay Studios".'

Their primary studio is in their Florida home, but between June and September they escape the heat and move to their summer studio in Blue Island, Illinois, just south of Chicago. This allows them to participate in outside art fairs in the Midwestern states, a primary source of their income. While in Florida, Glenn works full-time in the studio and teaches one night a week at the Dunedin Fine Art Center, Tampa. He also organises and facilitates at least one annual crystalline glaze workshop, using the Art Center as the host location.

Glenn and Keith use a custom blend clay made by Highwater Clays. It's a 50/50 mix of their Helios porcelain and Loafer's Glory white stoneware. Because they fire to cone 10/11, they initially found that the Helios clay alone bloated and slumped because it could not take the hotter temperatures. The blend with the Loafer's Glory works well and has eliminated the problem.

Glenn uses both crystalline and traditional glazes together. He begins by sketching each piece and indicating which glaze will go where, in what order, and what portion of the pot will be crystalline glazed. Glenn says:

> *'I like a colour blend from top to bottom, so I use three to four different colours; it could be the same crystalline glaze base with additions of colouring oxides, or it could be a variety of other bases to achieve this colour shift. Because the crystalline glaze flows so much during the firing, the colours often run together so that you don't notice the blending. It's difficult to choose a favourite glaze as we experiment with gloss versus matt, and oxidation versus reduction. My favourites are all over the board. However, this glaze has continued to be a studio favourite and is most reliable.'*

RIGHT Both © Glenn Woods.

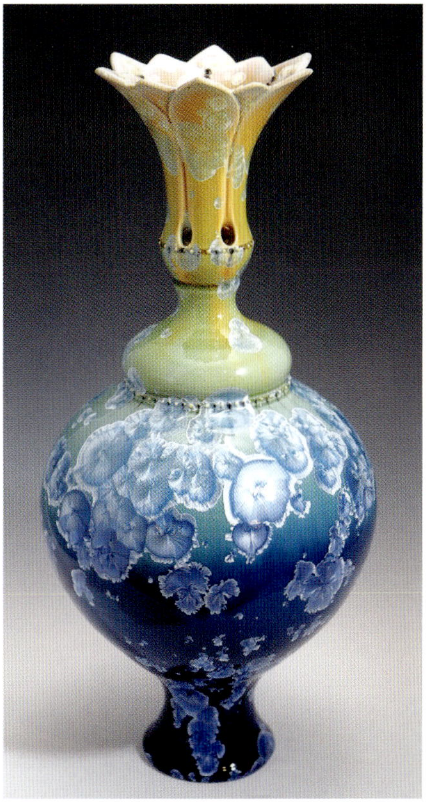

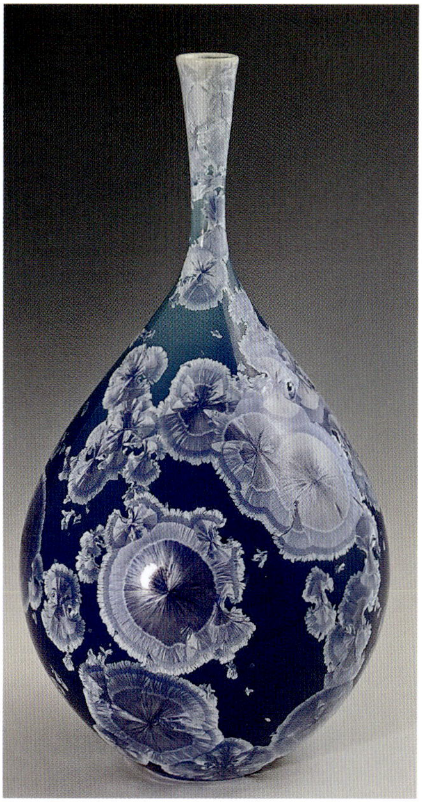

P11 Glaze: Cone 10/11

INGREDIENT	PARTS
Frit 3110	50.80
Zinc oxide	25.70
Silica	22.50
Alumina hydrate	1.00
Total	100.00
Add:	
Copper carbonate	2%
Cobalt carbonate	0.2%

Glenn and Keith have seven kilns in their Florida studio and four in Illinois, each dedicated to a specific task: bisque, gloss crystalline, high-fire reduction, post-fire reduction, and matt crystalline. This helps with achieving consistent glaze results. All their kilns are outfitted with controllers – some with Bluetooth technology – so they can monitor and make adjustments remotely.

They also have two styles of kiln for reduction firings. For post-fire reduction, they use a kiln custom-built for them by Jen-Ken Kilns of Florida; it looks like an electric kiln, but without grooved bricks, and it has a side port at the bottom for a propane torch. It also has a removable stack/chimney with a damper on top to increase draft and adjust the level of reduction.

Their second style is electric, altered to allow fuel to be dripped inside at the end of the crystalline firing. First, they fire to maximum temperature to grow the crystals, then drop the temperature to around 1050°C (1922°F) and reduce for 20 minutes using methanol.

Glenn has two main firing schedules for the P11 glaze, each giving a different result. The firing in Fig. 2 produces good-sized, free-form crystals surrounded by three more static halos, and one final halo that usually has a lacey pattern.

Fig. 2

RAMPS	TEMP °C	TEMP °F	HOLD
194°C (350°F)/hr	1285	2345	5 min
Ramp down	1149	2100	2.5 hr
28°C (50°F)/hr	1066	1950	45 min
33°C (60°F)/hr	982	1800	30 min
Ramp up	1010	1850	30 min
Ramp up	1104	2019	20 min

Fig. 3 shows a firing programme that produces more symmetrical, static crystals with smaller secondary crystals.

Fig. 3

RAMPS	TEMP °C	TEMP °F	HOLD
278°C (500°F)/hr	1274	2325	15 min
Ramp down	1093	2000	3 hr
Ramp down	1038	1900	45 min
Ramp down	1010	1850	30 min
Ramp down	982	1800	20 min
Ramp up	1010	1850	30 min
56°C (100°F)/hr	954	1750	20 min
33°C (60°F)/hr	760	1400	Off

Glenn says:

> *'I get a great deal of satisfaction from making well-crafted, beautifully formed vessels. Finding the right crystalline glaze coupled with the perfect firing schedule is pure magic. I also love teaching and sharing the joy of crystalline glazing with others.'*

Yadollah Ziarati, Iran

Yadi Ziarati resides in Tehran, the capital city of Iran; however, his pottery studio is in Semnan, 200 kilometres away. For four days a week, he works at his studio. For the remaining three, he returns to Tehran, where he holds crystalline glaze and low-fired porcelain workshops.

The studio in Semnan occupies 200 m^2 (2,153 sq ft). It's equipped with various tools for throwing, casting, and glazing. Yadi has a large, gas, shuttle kiln with a capacity of about 5 m^3 (177 cu ft), an electric kiln of 0.5 m^3 (18 cu ft), and another smaller electric kiln that he uses for experimenting with crystalline glazes. Yadi explains:

> *'The availability of materials here is of great concern to me. For instance, frit 3110 is impossible to obtain so I've developed my own equivalent, which I make available to my students and other interested potters. Additionally, I've formulated glazes to suit a firing schedule of up to 1200°C (2192°F) and I fire my macrocrystalline glazes at 1180°C (2156°F). The majority of potters in Iran have kilns capable of going only to 1100°C (2012°F). I often get asked if porcelain can be made at a lower temperature, such as 1100°C (2012°F). This became my main focus. Eventually, I succeeded in creating a low-fire porcelain. It has zero water absorption and is translucent.'*

Yadi is currently holding workshops and classes to teach this new porcelain formulation, which has been well-received by his fellow potters. He also makes his own stoneware, a combination of Iranian kaolin, feldspar, silica, and other additives.

Yadi has done his best to educate potters about crystalline and other complex glazes. He has written about crystalline glazes and made his research accessible to ceramicists. Yadi says:

> *'My main focus now is no longer on sales but on experimenting with new art forms. However, I have occasionally sold unique pieces to galleries and on social media. I make the majority of my living teaching and holding porcelain and glaze workshops.'*

Favourite recipe: Cone 6

This recipe gives a cream background with large blue crystals.

INGREDIENT	PARTS
F.Z.1 frit (can substitute frit 3110)	54.00
Zinc oxide	24.10
Silica	17.30
Boric acid	3.30
Lithium carbonate	1.10
Cobalt nitrate	1.00
Titanium oxide	7.00
Bentonite	2.00
Total	109.80

Firing schedule

RAMP	TEMP °C	TEMP °F	HOLD
1 hr to	120	248	0
2 hr to	460	860	0
4 hours to	960	1760	0
3 hours to	1220	2228	15 min
Fast drop to	1050	1922	4 hr

'Though I've put my blood and sweat into the studio and it serves as my workplace, I also enjoy the olive, pomegranate, grape, and fig trees that I've planted nearby. This is where I find my inspiration.'

BELOW Both © Yadollah Ziarati.

P.192 © John Tilton.

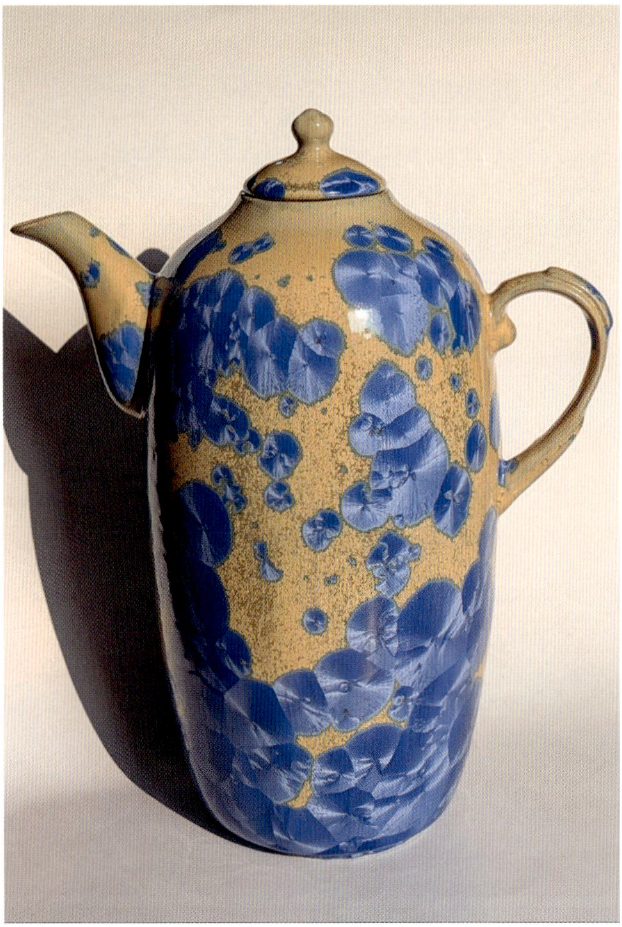

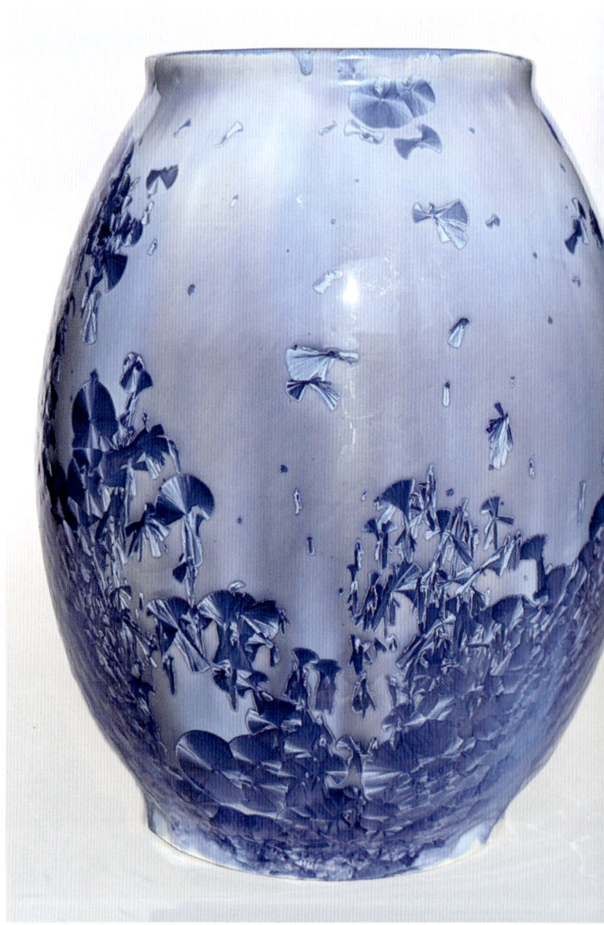

Bibliography

Alsop Robineau, Adelaide, 'Royal Copenhagen Porcelain', in *Keramic Studio*, vol. 3, May 1901–April 1902, pp. 10–11. Accessed at library.si.edu/digital-library/book/keramic-studio

Bodelsen, Merete, *The Royal Copenhagen Porcelain Manufactory 1775–1975*, The Manufactory, Copenhagen, 1975

Creber, Diane, *Crystalline Glazes*, Herbert Press, 2024

Evans, Paul, *Art Pottery of the United States: An Encyclopaedia of Producers and Their Marks*, Scribner, New York, 1974

Eidelberg, Martin, 'Robineau's Early Designs', in *Adelaide Alsop Robineau: Glory in Porcelain*, ed. Peg Weiss, Syracuse University Press, Syracuse, New York, 1981

Fäy-Hallé, Antoinette, *Porcelain of the Nineteenth Century*, trans. Barbara Mundt, Rizzoli International Publications, New York

Perry, Barbara, 'American Ceramics 1920–1950', in *American Ceramics: The Collection of the Everson Museum of Art*, ed. Barbara Perry, Rizzoli International, New York, 1989

Piché, Thomas Jr. and Julia A. Monti, eds., *Only an Artist: Adelaide Aslop Robineau, American Studio Potter*, Syracuse University Press, Syracuse, New York, 2006

Read, F. H., 'The University City Venture', in *Adelaide Alsop Robineau: Glory in Porcelain*, ed. Peg Weiss, Syracuse University Press, Syracuse, New York, 1981

Rhodes, Daniel, *Clay and Glazes for the Potter*, Chilton Book Company, Radnor, Pennsylvania, 1974

Sanders, Herbert H., *Glazes for Special Effects*, Watson-Guptill Publications, New York, 1974

Schmidt-Walguny, Axel, ed., *Kristallglasuren, Kreisverwaltung des Westerwaldkreises*, Montabaur, 1987

Snair, David, 'Making and Firing Crystalline Glazes', in *Ceramics Monthly* 23/10, 1975

Wakefield, Hugh, 'Modern Industrial Ceramics (from 1860)', in Robert J. Charleston, ed., *World Ceramics: Illustrated History*, Hamlyn, London, 1968

Acknowledgements

I am a potter first, so I thank those who read parts of my manuscript and offered suggestions and valuable criticisms, especially the women of my writing group: Nancy Bayley, Laurie Ness Gordon, and Zabe MacEachren. A big thanks goes to Bill Boyd who read the complete manuscript and offered pottery advice. I am also grateful to John Tilton and Richard Bideau for their valuable contributions as well as to reader Deb Twiddy. To the many potters who generously contributed to the book with information and images, I owe my deepest appreciation and admiration.

For technical expertise, I thank geochemist, Dr Gregory Lester, Jan Stoltmann at Rohde Industry, Dimitri Lewicki at L&L Kiln Mfg., Megan Vassar, Scotty McMillan and Cheyenne Terbrueggen at Skutt Ceramic Products, Sofia Boubakir at Geil Kilns, Frank Tucker and Michael Leonard at Tucker's Pottery Supplies and Molly Dishman and Tamara Bartlett at Bartlett Instrument Company.

I also wish to thank Garth Johnson, Curator, Everson Museum of Art, and Richard D. Mohr, Professor, University of Illinois.

I would also like to thank the staff at Bloomsbury Publishing (Jayne Parsons and Sara Simper) as well as my copyeditor, Rachael Prest and book designer, Laura Woussen.

Big thanks and hugs to my patient and talented husband, Tim De Rose: artist, potter, and best friend.

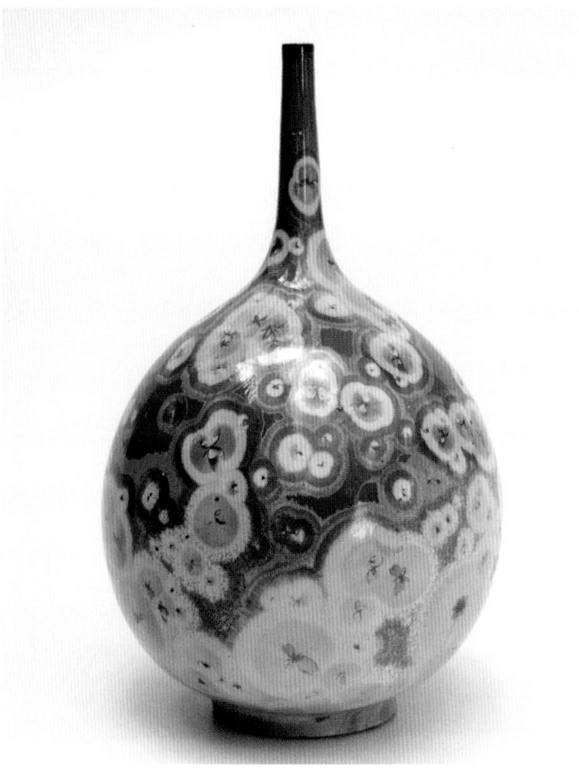

Index

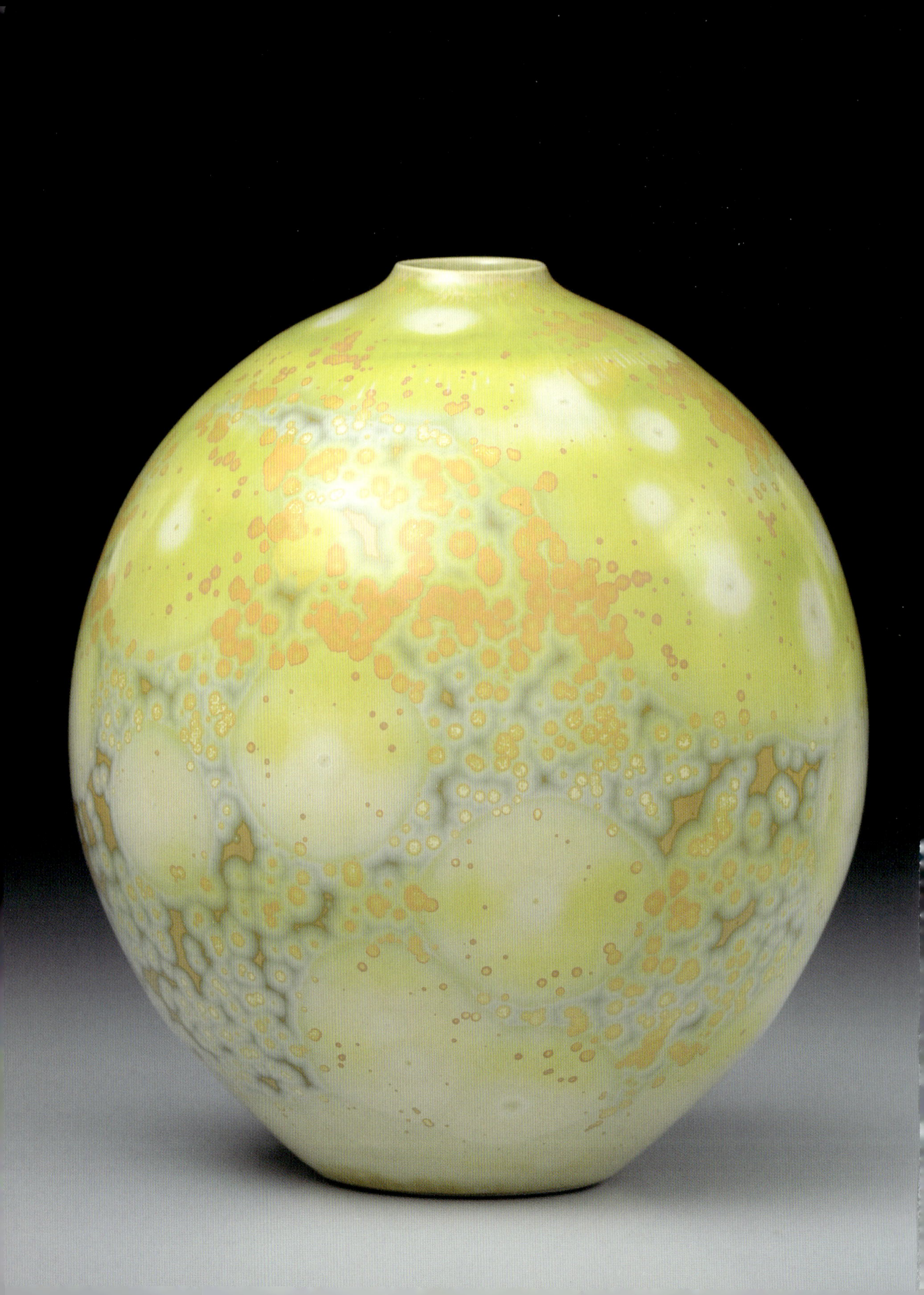